THE ENVIRONMENT OF MANKIND

AN INTRODUCTION
TO
PHYSICAL GEOGRAPHY

CARTOGRAPHY BY DAVID L. FULLER

HAMILTON PUBLISHING COMPANY
Santa Barbara, California

The Environment of Mankind

AN INTRODUCTION
TO
PHYSICAL GEOGRAPHY

Donald K. Fellows

CALIFORNIA STATE UNIVERSITY, NORTHRIDGE

Copyright © 1975, by John Wiley & Sons, Inc.

Published by **Hamilton Publishing Company**
a Division of John Wiley & Sons, Inc.
All rights reserved. Published simultaneously in Canada.

Library of Congress Cataloging in Publication Data:

Fellows, Donald Keith.
  The environment of mankind.

  Includes bibliographies and index.
  1. Physical geography. I. Title.

GB55.F36   551   74-32158
ISBN 0-471-25718-4

Printed in the United States of America

10 9 8 7 6 5 4 3 2 1

To Virginia

ABOUT THE AUTHOR   **Donald K. Fellows** received his bachelor's degree at the University of California at Los Angeles and his master's degree from California State University, Northridge. He is widely traveled, having covered much of the Pacific Ocean area during and after World War II; toured the Soviet Union, from European Russia to Siberia, Central Asia, and the region south of the Caucasus, primarily Soviet Georgia; Great Britain and Ireland; parts of Mexico and Canada; and most of the United States. His systematic interests in geography are in the human-cultural fields and their intimate relationship with the physical environment. He has written *A Mosaic of America's Ethnic Minorities* (published by John Wiley), a study of America's most visible minorities. In addition to teaching geography at California State University, Northridge, he teaches off-campus programs in geology and geography for local community colleges. He is a member of the Association of American Geographers, American Geographical Society, National Council for Geographic Education, Association of Pacific Coast Geographers, National Geographic Society, California Council for Geographic Education, and the Los Angeles Geographic Society.

This approach to the study of mankind's physical environment is not intended to be the final, analytical, theoretical, and all-inclusive answer. No attempt has been made to boggle the student's mind with intricate or overly technical terminology. Neither is the assumption made that the student is incapable of understanding scientific matters. Rather, the objective has been to explain the reality and the excitement of our planet in a geographic language that any nonprofessional can understand and enjoy.

This book is divided into four major parts, further subdivided into sixteen chapters, all of which endeavor to clarify the intimate relationship of the sun to the earth and its climates, vegetation, waters, soils, and its sometimes awe-inspiring surface features. Part 1, The Earth in Space, describes the true shape of the earth, how it is portrayed on globes and flat maps, and its role as a "spaceship" orbiting through the solar system powered by energy from the sun.

Part 2, Earth and Its Climates, focuses on the importance of the atmosphere, the differences in air temperatures and pressures across the face of the earth, local winds and wind systems, moisture, clouds, and storms, the earth's water supply, vegetation, and the differing climates of the world and their influence on human endeavors.

Part 3, The Earth's Structure, is a look at the earth's crust and its deep interior along with a consideration of the powerful forces that uplift mountains and continents and make the earth's land surface and sea bottom so interesting. Part 4, The Earth's Sculptured Face, explains how the dynamic, moving forces within the earth are countered on the surface by the downgrading, leveling processes of gravity, rain, and the different forms of running water, glaciers, oceans, and winds. The epilogue, The Environment of Mankind— Tomorrow?, offers some suggestions about how the information learned can be put to positive use in the effort to preserve our natural physical environment.

To further the learning process, there are several questions at the end of each chapter, based on the material covered. The wise student will attempt to answer them before proceeding to the next section. There are selected citations and suggested additional readings throughout the book, within or at the end of each chapter. These are intended for the serious student who desires to expand his knowledge beyond these pages.

I would like to express my appreciation and gratitude to those who have aided and encouraged me in this endeavor: Arnold Court, John Gaines, Eliot McIntire, and Charles Martinson, my colleagues in the Department of Geography, Northridge; Frank Hanna, Department of Geology, Northridge; Mark Powell, John Carthew, and Barry Haskell, Life and Earth Sciences Department, Los Angeles Pierce College; Robert Glendinning, Professor Emeritus, Department of Geography, University of California at Los Angeles, who long ago inspired me to enter the field of geography; and the following professors who gave much of their valuable time to review the several stages of the manuscript and offer invaluable advice: Clifford Tiedemann, University of Illinois, Chicago Circle; Rainer Erhart, Western Michigan University; Laura Copeland, Del Mar College, Corpus Christi, Texas; Merrill K. Ridd, University of Utah; Donald Kiernan, Solano Community College, Suisun City, California; and Richard Raskoff, Los Angeles Valley College, Van Nuys, California.

In addition I would like to express my thanks to various people associated with Hamilton Publishing Company: Paul A. Lee (Principal Editor), Ronald Q. Lewton (Editor for Production), Judith Chaffin (production and photographic editor), Janet D. Bollow (designer), David L. Fuller (cartographer and illustrator), Bonnie Harrington (copy editor), and Harm J. de Blij, Department of Geography, University of Miami, Coral Gables, Florida (Geography Advisor).

Donald Keith Fellows

CONTENTS

ix

THE EARTH IN SPACE

PART 1

1 On the Threshold...

He who does not rouse himself when it is time to rise, who, though young and strong is full of sloth, whose will and thought are weak, that lazy and idle man never finds the way to knowledge.

THE BUDDHIST DHAMMAPADA

There are three classes of people in the world. The first learn from their own experience—these are wise; the second learn from the experience of others—these are the happy; the third neither learn from their own experience nor the experience of others—these are fools.

LORD CHESTERFIELD

The more extensive a man's knowledge of what has been done, the greater will be his power of knowing what to do.

BENJAMIN DISRAELI

The objective of this book is to help you, the reader, to become more aware of life here on earth, to gain a greater understanding of yourself and your relationship to earth, and to comprehend the seriousness of the problems of pollution, over-population, and energy shortages. In considering the future prospects of mankind on earth, do you wonder what you might do to help build a better future? If, perhaps, you heed those words spoken by Disraeli and try to acquire the knowledge of what has been done to the earth by mankind, then, it is hoped you will be able to do something to help correct the wrongs.

There you have it: You are now on the threshold of a voyage of discovery, a voyage that will take you all over the face of this earth, an exciting journey that will enable you to form opinions based on fact, not emotion. You will stand in a much better position than the ancients who devised theories out of dreams and "visions," rather than from experience.

Early Theories of Earth's Systems

It has been a relatively short span of time—perhaps a few thousand years—during which human life has been somewhat secure, when people could sit and ponder the complexities of their existence. They were then able to consider the earth under and about them and the great universe beyond. They evolved a number of theories and ideas to attempt to explain their origin and their place in the scheme of things. Many of these people were profound thinkers, and they came up with answers that were satisfying for a time. One of these, the Chinese sage Kung Fu-tze (Confucius), said that the four seasons pursue their course without the aid of man, and all things are constantly being produced. He thought that the spirits of heat and cold, sun, wind, rain, stars, forests, streams, mountains, and valleys could be made happy and won over to man's side by offering sacrifices.[1] Another Chinese, Chuang Tze, also pondered the mysteries of the sky and wrote:

The sky turns round; the earth stands still; sun and moon pursue one another. Who causes this? Who directs this? Who has leisure enough to see that such movements occur? Some think there is a mechanical arrangement which makes these bodies move as they do. Others think that they revolve without being able to stop. The clouds cause rain; rain causes clouds.[2]

And one of the great Jewish prophets, the second Isaiah, wrote:

Who hath measured the waters in the hollow of his hand, and ruled off the heavens with a span, and inclosed the dust of the earth in a measure, and weighed the mountains with a balance and the hills in scales?[3]

In those days, there was insufficient knowledge to discover the reasons behind the changing of the seasons, the transition from day into night and night into day, the formation of stream valleys, or the beneficent or detrimental effect of the environment upon mankind. Rather, all these phenomena were regarded as the work of nature spirits or gods, supernatural beings such as Zeus of the Greeks, Shang-ti of the Chinese, Dyaus Pitar of the horsemen of the steppes of Central Asia, Ahura Mazda of the Persians, Yahweh of the Hebrews, or Odin of the Scandinavians.

FROM METAPHYSICAL TO PHYSICAL THEORIES   Change is inevitable, however, and in later times new ideas arose as men tried to come up with better answers. Consider the ancient Greek philosophers, such as Homer (ninth century B.C.), who thought the earth was a flat disk surrounded by the river "Ocean"; Anaximander (sixth century B.C.), who thought the earth was a giant cylinder hanging suspended in space; and Pythagoras (sixth century B.C.), who advanced the purely philosophical theory that the earth was neither flat nor cylindrical, but was a sphere, a round, ball-shaped bit of matter floating in space.

None of these men concerned themselves with the question of the actual origin or creation of the earth; they were interested only in explaining the various phenomena they witnessed, using the best logic they could muster. Their answers were satisfying to many, for those were times when men traveled afar but understood little of the true vastness of the earth. They wrote extensively of their travels, but most of these early writers merely described what they saw, and their observations were colored by the basic knowledge of the time.*

It might be said that today we do the same thing, except that we have more data, more statistical information, better scientific equipment, and considerably more knowledge and experience to compare or contrast the reasons for the differences and similarities in the earth's phenomena. The thinkers before our time should not be laughed at or blamed for coming up with erroneous conclusions, any more than we should today.

ENVIRONMENTAL DETERMINISM   Another of the eminent Greeks, the noted metaphysician and philosopher Aristotle (fourth century B.C.), attempted to explain why the weather and climate varied from one place to another in the world he knew. He divided the earth into three regions of climate, or **klimata** : (1) the torrid zone, which included the Great Desert (Sahara) and the grasslands to the south, (2) the frigid zone to the north, a land of forests and snow, and (3) in between, the temperate zone, the region surrounding the Mediterranean Sea. This was a consid-

---

* For an interesting accounting of these courageous explorers, see: M. Cary and E. H. Warmington (1963), *The Ancient Explorers* (Baltimore, Md.: Penguin Books). See also: E. H. Bunbury (1959), *A History of Ancient Geography,* vols. 1 and 2 (New York: Dover Publications).

erable step forward for the time. Aristotle's thinking reflected the extensive explorations the Greeks were making into Africa, Asia, and Europe. His position was similar to our own today with relation to our explorations into outer space. We do not scoff at Aristotle's *klimata,* but we do shake our heads at his interpretation of the influence of climate on the affairs of mankind, particularly as it might be applied to present-day society's technical or cultural achievements. Aristotle rationalized that the "white-skinned barbarians" who lived in the cold, snowy frigid zone would never amount to much, because the climate in northern Europe was too severe. As for the torrid zone, he believed that the "black-skinned savages" should not be expected to be very progressive, because life would ever be too easy for them. They would be doomed for all time to wallow in idleness. Aristotle concluded that Greece, because of the ideal climate of the Mediterranean Sea, was the center of the universe. He believed that the "olive-skinned peoples" of the Mediterranean region were obviously destined to rule the world or, at least, to show the way to all other races of mankind.

One might well smile over Aristotle's simplistic analysis, but his fourth century B.C. reasoning persisted through the nineteenth century. Indeed, the concept of **environmental determinism** grew through time, as other respected scholars reinforced his theory: Charles Darwin and Sir Charles Lyell (England), Friedrich Ratzel (Germany), Ellsworth Huntington and Ellen Churchill Semple (United States), among others.* Aristotle's theory was shaken by the growing awareness of the existence of great civilizations and outstanding cultures that evolved in both the frigid and torrid zones; the Vikings, Germans, Russians, and English in the frigid zone, and the Inca of Peru, the Maya of Yucatan, as well as the empires of Ghana, Mali, and Songhai of Africa in the so-called torrid zone.

The twentieth century, however, brought the theories of environmental determinism to an end. Most geographers of today, along with other theoreticians in the earth sciences (Table 1–1), accept the idea that the physical environment does exert significant influence, but they do not believe that it controls mankind's destiny or direction. Other mammals may be forced to adapt to their environment, but humans have free choice. If a particular society does not like the weather or climate where it resides, it can move. Or, it can alter its climate by modifying its structures. Today, we use insulation, put trees around our homes, and install heating or air-conditioning units. In other words humans can—within limits—either choose to adapt to their environment, leave it, or make it conform to their own desires.

Mankind and Nature

Perhaps the very freedom to control their environment has brought modern societies to the predicament in which they find themselves—a world becoming dangerously polluted. In general, we might point the finger at the technological advances of

---

*For an interesting discussion of the various concepts of environmental determinism, with reference to the influence or direct control of the physical environment on mankind, see: A. H. Meyer and J. H. Streitelmeir (1963), *Geography in World Society: A Conceptual Approach* (Philadelphia, Pa., and New York: Lippincott), chap. 1, pp. 2–25.

TABLE 1-1   The earth sciences

THE EARTH SCIENCES	THEIR OBJECTIVES
GEODESY	Investigations to determine the true shape of the earth.
PHYSICAL OCEANOGRAPHY	Study of the oceans, their currents, and the ocean floor.
ASTRONOMY	Study of the relationship of the earth to the sun, moon, and stars.
CLIMATOLOGY	Examination of the effects, over long periods of time, of solar radiation—sun energy—on the earth and the physical environment.
METEOROLOGY	Study of weather—that is, short-term atmospheric conditions, day-to-day variations in climate.
HYDROLOGY	Study of earth's freshwater supply: streams, rivers, lakes, groundwater.
PLANT GEOGRAPHY	Investigation of the vegetative cover of the earth: desert shrubs, grasslands, forests.
PEDOLOGY	Study of the soils of the earth.
PETROLOGY	Study of the rocks of the earth.
MINEROLOGY	Study of the minerals of the earth.
GEOLOGY	Examination of the structure of the earth.
GEOMORPHOLOGY	Examination of the face of the earth: its landforms.
CARTOGRAPHY	Presentation, by means of charts and maps and graphs, of information gleaned from any source.
GEOGRAPHY	Analysis of the features of the earth: their location and distribution and how they are related to each other.

Geodesy

Physical Oceanography

Astronomy

Climatology

Meteorology

Hydrology

Plant Geography

Pedology

Petrology

Mineralogy

Geology

Geomorphology

"Western culture," for not all culture groups believe that humans rate above nature. Consider, for example, the nomadic Masai of Kenya and Tanzania in East Africa. They still believe that man and animal—and the soils and vegetation as well—are one. Humans are not to do as they please with other living forms or the earth itself. In their opinion, humans are, at best, caretakers, protectors.

And far to the northeast, the Hindus of three or four thousand years ago came up with this hymn to the oneness of life:

All hail to heaven! All hail to earth!
All hail to air! All hail to air!
All hail to heaven! All hail to earth!
Mine eye is sun and my breath is wind, air is my soul
 and earth my body.
I verily who never have been conquered give up my life to heaven
 and earth for keeping.[4]

It would appear that the Eastern religions have placed humans not on pedestals but directly *in* nature, an integral part of nature. Traditionally, ancient thinkers of eastern Asia, such as Gautama the Buddha and Confucius, held a strong ecological approach to humans and their relationship to their environment. It is quite possible, however, that the tremendous increase in the population of Asia during the past century will shake this tradition and bring about a greater despoilation of their once-sacred earth.

One must also be wary of placing too much blame on the Western peoples. Perhaps some have taken the words of the Bible's Genesis 1:26 too literally:

Let us make man in our image, after our likeness, and let him have dominion over the fish of the sea, the birds of the air, the domestic animals, the wild beasts, and all the land reptiles.

Perhaps, also, the violators of our earth should have read further, to Genesis 2:15, and heeded these words:

Then the Lord God took the man and put him in the garden of Eden to till it and look after it.

If the latter statement had been stressed as strongly as the first, mankind might not have decimated the forests, the birds, the beasts of the fields and seas.

To reverse this trend, to bring about a greater ecological awareness, we must know more about the earth on which we reside. We must realize that we are not *above* nature, but are an integral part of it. It is to be hoped that we can learn and understand what might be termed a *land ethic*—a realization that the earth is not just a toy to be used and thrown away. The earth—its waters, soils, vegetation, and landforms—as well as our atmosphere, is mankind's home, our place of residence. In reality, it is not the environment so much as it is the manner in which we use it that counts.

The Geographic Study of the Earth

To truly understand this earth of ours, we need to know more about a lot of subjects. The geographer is trained to look at the world in a highly comprehensive way. He uses the work of many specialists (see Table 1–1) in his research. He consults pedologists (soil scientists) to learn about soils; he studies the work of botanists, geologists, zoologists, agronomists; and he collates the compiled material. To understand the effect of the sun and moon upon our planet and its waters, he is concerned with astronomy. To understand the movements and the work of oceans, he consults the oceanographer. He uses all the help he can get so that his answers to the earth's problems will have objectivity and validity. The physical geographer must try to understand how all the different elements of our earth fit together—how they mesh, relate, and affect each other.

The geographer wants to know why people, places, and things differ from place to place. He is interested in the *location* of things—not merely that they exist in whatever form, but *where* they are and *why* they exist where they do. He wants to explain the distribution of people, places, and things. Physical geographers seek the reasons for the distribution of volcanoes, fjords, tornadoes and hurricanes, forests and grasslands. Human geographers take all this information into consideration and try to understand the relationship of the physical elements to the distribution of human population (Fig. 1–1) across the face of the earth.

Some geographers prefer to approach their investigation and study by dividing the earth up into sections, or regions, then making a thorough study of a particular region's climate, soils, vegetation, landforms, religions, political preferences, industrial activities, and so on. This is termed the **regional** approach. Others, however, prefer to consider the physical or cultural elements system-by-system all around the earth, which is termed the **systematic,** or topical, approach. In this manner, the systematic geographer comes to understand the spatial interaction of all the elements. For example, a geographer using the systematic approach to investigate the wind systems of the earth would study the influence of the sun in causing high- and low-pressure zones, the movement of air from one place to another, and the influence of the rotating earth on wind direction. He would also be concerned with how the wind in an area affects the overall climate, rainfall, fog conditions, and vegetation. Are there any similarities between California and Illinois, or between Moscow, U.S.S.R., and Moscow, Idaho? Some geographers want to step back and take the broad world view, and some want to zero in on small-area complexities.

Here, we shall look at the earth system-by-system, feature-by-feature, physical-element-by-physical-element, and we shall try to show how all the different physical elements fit together—and what effect they have on different peoples' selections of places of habitation. In order, we shall consider the earth itself: its shape; how it is portrayed on globes and maps; its weather and climate; its waters, vegetation, and soils; its structure and composition; the dynamic forces at work beneath the surface and above it that have given our earth its fantastic and beautiful sculptured face.

And when we have finished, you will have a better understanding of your earth, your home.

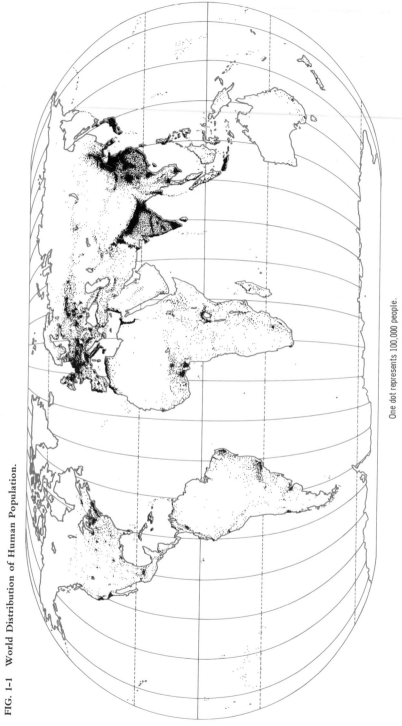

FIG. 1-1 World Distribution of Human Population.

One dot represents 100,000 people.

Review and Discussion

1. From your personal experience and background, how would you compare the approaches of the ancient geographers to those of today?

2. Discuss Aristotle's theory of the *klimata*. Is there any validity to his theories on the division of the earth into three climatic zones?

3. After doing some outside reading of the works of Ellen Churchill Semple and Ellsworth Huntington, discuss their concept of environmentalism.

4. What do you think of the idea that "man is an integral part of nature"?

5. Explain the importance of the geographic concept of location.

6. Explain the difference between the regional and systematic approaches to the geographic study of the earth.

7. Compare and contrast the various earth sciences: geodesy, oceanography, astronomy, climatology, meteorology, hydrology, pedology, geology, geomorphology.

Notes

[1] Robert O. Ballou, ed. (1967), *The Portable World Bible* (New York: Viking Press), pp. 486–490.

[2] Ballou, p. 561.

[3] Isaiah 40:12–13.

[4] Ballou, p. 37.

Additional Reading

On the history of geography, its character, its essence, and its evolution from ancient times to the present, see:

Bunbury, E. H. (1959). *A History of Geography*. Vols. 1 and 2. New York: Dover Publications.

Broek, Jan O. M. (1966). *Compass of Geography*. Columbus, Ohio: Charles E. Merrill.

Carter, George F. (1968). *Man and the Land*. 2d ed. New York: Holt, Rinehart and Winston. See, especially, the introduction.

Cary, M., and E. H. Warmington (1963). *The Ancient Explorers*. Baltimore, Md.: Penguin Books.

Meyer, A. H., and J. H. Streitelmeier (1963). *Geography in World Society: A Conceptual Approach*, chap. 1. Philadelphia, Pa., and New York: Lippincott.

Semple, Ellen Churchill (1911). *Influences of Geographic Environment*. New York: Henry Holt.

Warntz, William, and Peter Wolff (1971). *Breakthroughs in Geography*. New York: New American Library. Contains "the key writings of the great historical figures of geography."

2

PORTRAYING THE EARTH

Since the beginning of mankind's existence on the planet earth, individuals have attempted to portray the land about them. Some have done it verbally, with songs, chants, and poems as they described the directions, distances, and landmarks for their important routes. Others tied sticks together to form a layout of the region about them, or they inscribed lines in a mass of wet clay. These methods worked fairly well for those who knew only that bit of earth space about them, but think of the problems that arose when knowledge of the actual size and shape of the earth began to enter their understanding.

It was a relatively simple matter for our ancestors to construct maps of the vicinity of their homes or hunting grounds, but it became quite a different story when people became concerned with extensive areas and vast distances. Before maps of a continent or the earth itself could be constructed, a series of questions had to be answered, such as the size and shape of the earth, and how to make, without distortion, a flat representation of a round globe. The first mapmakers had to devise ways to pinpoint the location of places, to figure true sailing courses across large oceans, and to make the map readable and understandable to any who might want to use it.

Today we feel quite satisfied with our solutions, but it took a long time to find them. Humans tried, for thousands of years, to unravel the puzzles.

Earth's Shape: Ancient Concepts

Some 2000 years before the beginning of the Christian era, the Hindus compiled a book of Indian scriptures from which we learn:

> In the beginning, this was nonexistent. It became existent, it grew. It turned into an egg. The egg lay for the time of a year. The egg broke open. The two halves were one of silver, the other of gold. The silver one became the earth; the golden one, the sky; the thick membrane of the yoke, the mist with the clouds; the small veins, the rivers; the fluid, the sea.[1]

Centuries later, halfway around the world, an Apache of the American Southwest believed that in the beginning there was no earth—no world—there was nothing but Darkness, Water, and Cyclone—but:

> there were the gods who had always existed and who had the material out of which everything was created. They made the world first, the earth, the under-world, then they made the sky. They made the earth in the form of a living woman and called her Mother. They made the sky in the form of a man and called him Father. He faces downward and the woman faces up. He is our Father and the woman is our Mother.[2]

Several thousand years ago, the peoples of the island of Crete and the nomadic horsemen of the steppes of Central Asia held similar views, although the latter did emphasize the masculine and tended to downgrade the importance of the earth as a "mother." The Egyptians thought differently. They conceived of the earth as masculine, a recumbent earth-god, Geb, from whose body sprouted all the world's vegetation. Above him was the sky-goddess, Nut, *whose long, curved body touches the earth only with the tops of her toes and fingers. It was the starry belly of the goddess which men saw shining in the night above them.*[3]

The number of theories that try to explain the origin of the earth and its physical shape are as varied as the peoples who devised them. Homer, the illustrious Greek writer of the ninth century B.C., thought of the earth as a kind of a raft floating in space, surrounded by a mighty river he called *Ocean*. Tahitians of the South Pacific thought of the earth as being a giant tortoise swimming leisurely through space. Babylonians of the Middle East believed the earth was a hollow mountain around which an ocean swirled, and the sky was like a curved ceiling through which the stars and sun peeked. And almost all of these ancient theories considered the earth as fixed in a permanent position in space, while the sun, moon, planets, and stars marched overhead in a regular procession.

Earth as a Sphere

With the passage of time, people began to believe that the earth was not merely flat like a plate, but instead, had curvature. Imagine a person standing atop a hill, peering seaward toward an offshore island some 50 miles (80 kilometers) away. The

Ancient Egyptian Concept of Heaven and Earth. The goddess Nut's starry body forms the celestial "vault of heaven," and over her form sails the sun-god Ra once each day. Beneath Nut lies the earth-god Geb, whose body forms the earth's crust.

FIG. 2-1 Curvature of the Earth. Line A-D represents the line of sight of a person stand-ing on a hilltop looking toward an island. Line B-C is the line of sight of a swimmer. If the island were 50 miles (80 kilometers) distant, with an elevation of 1320 feet (400 meters), it would be clearly visible to the person at point A. The swimmer at point B, however, cannot see the island because the earth makes such a curve at that distance. In fact, the island would have to have an elevation of at least 1650 feet (500 meters) to reach the swimmer's line of sight.

person, deciding that the island might be easy to defend against attack by hostile neighbors, descends the hill and begins to swim out to the island. But when he looks up to get his bearing, he discovers to his dismay that the island has disappeared. Had some capricious god or spirit of the sea snatched the island away? Returning to his hilltop, he is aghast: There is the island. His confusion may have then given way to a new thought: The earth's surface must curve. He could see it only when he was high enough for his line of sight to encompass the island as it rose above the curvature. Figure 2–1 offers an explanation as to how such an event could occur. Of course, the earth's curvature is understood today by anyone who has seen photographs of the earth taken from satellites in orbit.

As the centuries passed, new theories evolved. Pythagoras (sixth century B.C.) philosophized that the only perfect shape in the universe was a sphere. As the earth (in his mind) was the center of the universe and, therefore, perfect, it *had* to be a sphere. This concept became common knowledge among the early Greeks and Romans. However, the first person of record actually to postulate some earth measurements was Eratosthenes of Alexandria, Egypt, during the third century B.C. Determining angles of noonday shadows cast by the midsummer sun at two different places along the Nile River, and measuring the land distance between the two places, he estimated that the arc, or angle, between these two places from the center of the earth was about one-fiftieth of a complete circle. Assuming that the earth was a sphere, he multiplied the distance by 50 and came up with a figure of the earth's circumference that was a little over 25,000 miles (40,000 kilometers). Although he did not have accurate measurements with which to work, his errors counteracted each other, and he arrived at this figure, which was within a hundred miles of today's accurately figured circumference of 24,902 miles (39,843 kilometers).*

* The following two treatments of the work of Eratosthenes are recommended: David Greenhood (1964), *Mapping* (Chicago, Ill.: University of Chicago Press), pp. 39–40; Desmond King-Hele (October 1967), "The Shape of the Earth," *Scientific American,* 217, no. 4, 4 (offprint no. 873).

The Curvature of the Earth. This photograph taken from *Apollo 9* clearly shows the curvature of the earth. In the center is the Strait of Gibraltar, the passage between Morocco and Spain, which connects the Atlantic Ocean and the Mediterranean Sea.

By the second century of the Christian era, the scientific study of the shape of the earth was advanced by two important men of history: Crates and Ptolemy. The philosopher Crates constructed a globe 10 feet (2.99 meters) diameter that, owing to the lack of adequate and accurate data, did not portray the landmasses and oceans of the world as we know them to be today. He was soon followed by a geographer-turned-mapmaker named Ptolemy, who put his specialties of astronomy and geography to work to prove that the earth was a sphere. He, too, constructed a globe and from it developed a map projection in an attempt to create a flat representation of his theoretical round earth. He used lines representing parallels of latitude and meridians of longitude (that had been developed some seven centuries earlier by Herodotus*) to locate places on his map of the known world.

Although Ptolemy's map carried forward the errors long held by geographers and navigators that the earth was considerably wider from east to west than it was from north to south, it was used for more than a thousand years.† Had it not been for his error in assuming that Spain and Asia were closer than they really were, it is possible that Christopher Columbus might never have set out on his epochal journey to "far Cathay" (China). Although Columbus did not reach the Indies, and the people he met were not "Indians," he did add further proof to the theory that the earth was round.

* Herodotus (fifth century B.C.) believed the earth was wider in its east-west dimensions than it was from north to south. He developed terms such as *longitude* (for the longer east-west directions) and *latitude* (for the lateral, or lesser, north-south directions). His thinking affected map making for centuries.
† For an interesting look at the actual words used by Ptolemy to explain his theories, see William Warntz and Peter Wolff (1971), "Ptolemy: The Classical Synthesis," *Breakthroughs in Geography* (New York: New American Library), chap. 1, pp. 15–60.

Ptolemy's Map of the World—Second Century A.D. (Norman J. W. Thrower, *Maps & Man: An Examination of Cartography in Relation to Culture and Civilization,* © 1972, p. 45. Reprinted by permission of Prentice-Hall, Inc., Englewood Cliffs, New Jersey.)

Earth from Space. December 1972; and *Apollo 17* flashes toward the moon. The members of the crew look back to see the dazzling interaction of the atmosphere with the rays of the sun: swirling clouds spiraling out from Antarctica, and Africa, a not-so-dark continent.

Old beliefs, ideas, and traditions die hard, however, and there are still some people who will not accept the "round earth" doctrine. Today in England, for example, there is a small group of supposedly well-informed people who have formed the Flat Earth Society of Dover. Despite a dwindling membership, these staunch believers in the flat-earth theory scoff at our so-called proofs. They verbally attack America's space accomplishments as merely fraudulent attempts to deceive the people and to promote the vested interests of manufacturers of maps and globes. As they cling to the ancient belief that the earth is the center of the universe, they dismiss television broadcasts and satellite photographs from outer space as "cinematic trick photography made in Hollywood."[4]

With all due respect for the rights of people to believe as they will, there is no need for deception. The earth is round—that is, it has a roundish shape, slightly elliptical (oval-shaped) in form (Fig. 2–2). The distance from one side of the earth to the other, measured through the equator, is about 7926 miles (12,681 kilometers). The distance from the North Pole to the South Pole, measured through the center of the earth, is about 27 miles (43 kilometers) shorter. Its form is that of an ellipse, but one that is squashed in a little at the poles and a little fatter about its equatorial

FIG. 2-2 Shape of the Earth. Among these highly exaggerated sketches, (a) is a true sphere, and the distance from A to A' is the same as that from B to B'. (b) shows the distortion from the true sphere caused by the earth's rotation: The earth bulges slightly around the middle; thus, the distance from B to B' is greater than that from A to A'. (c) is a geoid. Although not to true scale, it is representative of the earth's elliptical shape and the mountainous highs and oceanic lows of the earth's surface. (d) gives the dimensions from pole to pole (A to A') and through the earth at the equator (B to B').

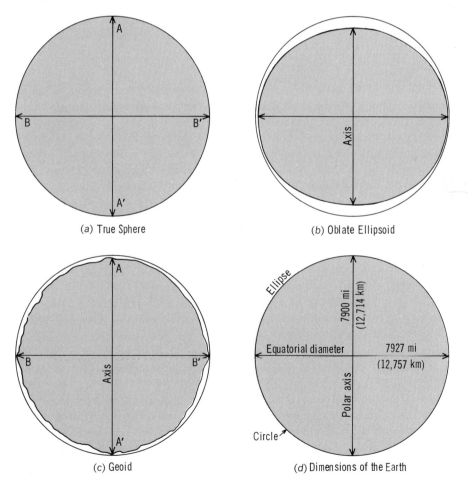

(a) True Sphere

(b) Oblate Ellipsoid

(c) Geoid

(d) Dimensions of the Earth

middle. The reason for the bulge around the equator is the earth's rotation: The spinning about the polar axis causes a displacement of the earth's bulk and a subsequent downpressing of the polar regions. The earth's shape, therefore, is that of an **oblate ellipsoid.**

A closer look at the earth also tells us that it does not have the smooth shape of an elliptical bowling ball. It has, instead, an uneven surface: The elevations of the landmasses are much higher than those of the ocean floors, and in addition, there are great heights, such as Mt. Everest, towering to an elevation above sea level some 29,028 feet (8708 meters), as well as great depths in the oceans such as the Mariana Trench, which is 35,800 feet (10,740 meters) below sea level. Obviously, therefore, the earth has an uneven surface. Because of this, the **geodesists** (scientists who try to determine the exact shape of the earth) call the earth's shape a **geoid,** meaning "earth shape." This implies that our earth has a unique shape, apparently unmatched anywhere else in our solar system. Certainly we could not describe the shape of the planet Mars as that of a geoid, for Mars has its own distinctive shape and form.

While some earth scientists may require precise mathematical measurements in trying to accurately assess the size and shape of the earth, it is not necessary for us to concern ourselves with such miniscule differences. For our purposes, we can assume that the earth is a sphere.

THE GEOGRAPHIC GRID One of the things a geographer needs to do is to define location. Although we can rely on names or descriptions—street addresses, zip codes, names of cities, or verbal directions—a more accurate way is to use the **geographic grid.** Simply stated, this is a system of locating places on the globe or a map by using lines running east-west and north-south. These lines are parallels of latitude and meridians of longitude, and in order to understand these devices, we must first comprehend the difference between great circles and small circles.

Great circles and small circles If you were to take an orange, which is a sphere, and cut it in half, the resulting two pieces would be equal in size. The circumference of the exposed face of either half would be a complete circle, called a **great circle.** If you were to take another orange and cut it into two pieces, but your knife did not slice through the center of the orange, you would have two pieces, but they would be of different sizes. The circle representing the cut face would be a complete circle, but it would not be a great circle; it would be a **small circle.** Great circles always divide a sphere into two equal parts; small circles do not. All circles drawn on the face of a globe (which represents the earth) that have as their center the exact center of the earth, or those passing through both of the poles, are great circles. Consider Fig. 2–3. Sketch (*a*) displays a number of great circles drawn on a globe. All of these lines, or circles, on the globe have as their center the center of the earth. In sketch (*b*), only two great circles are shown, but these are the two most important ones. The north-south great circle is composed of two meridians (a meridian is half of a great circle), one being the **prime meridian** (or zero meridian). The other meridian would be 180 degrees away, on the other side of the earth. The east-west great circle in this sketch is the equator. Sketch (*c*) has a number of circles, but only

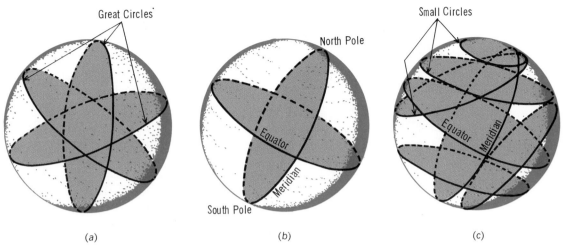

FIG. 2-3 Great Circles and Small Circles.

two are great circles. The others are small circles, and it is obvious that their centers are not at the center of the earth.

This concept of great and small circles is most important in navigation, whether for ships or planes, for the shortest distance between two different locations on the earth's surface is always along a great circle connecting the two points. Ship's courses are generally referred to as "great circle routes," and today's jets, flying from Los Angeles to London, usually fly a great circle route that takes them over the southern tip of Greenland. If a jet flew over New York en route to London, the distance would be much greater, for such a path would be along a small circle.

Meridians of longitude and parallels of latitude **Meridians of longitude** are halves of great circles. As the circumference of any circle is equal to 360 degrees, a meridian is 180 degrees in length, stretching from pole to pole (Fig. 2-4). If you look at a globe, you will note that all the meridians run in a true north-south direction. Spaces between meridians along the equator are greater than those along parallels of latitude away from the equator, for the simple reason that meridians converge at the poles. Any number of meridians can be drawn on a globe, and every place on the earth's surface has its own meridian. On most globes, however, the meridians are spaced about 10 or 15 degrees apart, and the first meridian, the prime meridian, which runs through the Royal Observatory at Greenwich near London, England, is the starting place. Figure 2-4(*b*) shows how meridians of longitude are determined, and you can see that longitude is the angular distance measured east or west of the prime meridian. Meridians to the west of Greenwich are referred to as meridians of west longitude; to the east, east longitude. When the angular count reaches 180 degrees east or west, the count stops. This is the 180-degree meridian, also known as the **International Date Line.***

* For clarification of the role of the 180-degree meridian as the International Date Line, see Appendix A, on time zones.

(a)

(b)

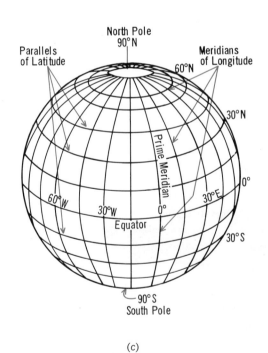

(c)

FIG. 2-4   Meridians of Longitude and Parallels of Latitude.

Parallels of latitude are the other lines used in determining the geographic grid. Knowing an area's meridian of longitude is only half the solution. We need to know just where along a meridian the area is located. We do this by finding which parallel of latitude crosses the meridian at that place. Parallels of latitude (see Fig. 2–4(*a*), 2–4(*c*) again) are small circles (except for the equator), and they are always parallel to the equator and to each other. They run in a true east-west direction, and they always intersect meridians at right angles (except at the poles, where there are no parallels). Any number of parallels can be drawn on a globe, but again, most globes have them spaced at 10 or 15 degree intervals, starting at the equator, the 0 degrees parallel, and ending up at 90 degrees north or south (the poles).

Therefore, to find an area's geographic grid position, we must determine both its latitude and its longitude. Where the lines intersect is the area's geographic grid position on the earth. Another important point to remember is this: Merely knowing the degree of latitude and longitude is not enough. We must also know whether the latitude is north or south and whether the longitude is east or west. Consider, for example, the geographic grid for Puerto Rico. It will not suffice to say "latitude 18°, longitude 67°," for there are four places on earth that have such a position. We must go further and complete the geographic grid this way: "Puerto Rico's geographic grid position is latitude 18° north, longitude 67° west." Without the north and west directions, you would not know where Puerto Rico is located. Consider the example illustrated in Fig. 2–5, which locates four positions that all have the same latitude and longitude (latitude 22°, longitude 79°).

Determining latitudinal and longitudinal position It is easy to pick up a map and point to your latitude—if you know where you are. But what if you are lost? You would then need an instrument called a **sextant.** This device is used to measure the angle between the horizon and a celestial body, such as the sun or the North Star

The Use of a Sextant. NROTC candidates at Prairie View A & M College, Texas, receive practical training in using the sextant.

FIG. 2-5 Using the Geographic Grid to Locate Places. Place A is at latitude 22° north, longitude 79° west and is in Cuba. Place B is at latitude 22° north, longitude 79° east and is in India. Place C is at latitude 22° south, longitude 79° east and is between Africa and Australia in the Indian Ocean. Place D is at latitude 22° south, longitude 79° west and is off the coast of northern Chile. What are the geographic grid positions of places W, X, Y, and Z?

(Polaris). For example, if you are in the Northern Hemisphere, taking a position facing north, you sight through the eyepiece of the sextant and measure the angle between the horizon and Polaris. That horizon angle is your latitude. Polaris can be used as a sighting point only in the Northern Hemisphere, however, because it cannot be seen south of the equator. If you are on the equator, Polaris appears to be resting on the horizon. As you move northward, Polaris rises 1 degree in angle for every degree north you move. For example, if you stopped along the way and found that your horizon angle was 60 degrees, your latitude would be 60° north. At the North Pole, the angle would be 90 degrees, since Polaris would be directly overhead.

Determining your longitude is somewhat less complicated. You do not need a sextant, but you do require two precise clocks (called *chronometers*) for telling the time. One chronometer tells you what time it is in Greenwich, England (through which the prime meridian passes), and the other tells you the local time. Because the earth rotates around its polar axis once every 24 hours, it moves from west to east 15 degrees every hour (360 degrees divided by 24 hours equals 15 degrees). If you know the time at Greenwich, as well as your local time, all you need do to determine your longitudinal position is to calculate the number of hours difference (Fig. 2–6) and multiply that number by 15. If your time is earlier in the day, you will be west of Greenwich; if your time is later than it is on the prime meridian, your longitude will be east. Further, if you keep in mind that international time is counted from 0000 (midnight) to 1200 (noon) to 2400 (midnight again), it is easier to figure the difference. Consider this: If your local time is 0900 (9:00 A.M.) and it is 1200 hours at Greenwich, your position is three hours earlier in the day, or 45 degrees west of Greenwich, and your longitudinal position is 45° west. One-half hour represents 7 degrees, 30 minutes. Therefore, if your time was 1430 hours, you would be two and one-half hours east of Greenwich, or 37 degrees, 30 minutes; thus, your longitudinal position would be 37°30′ east.*

EARTH'S SHAPE AND DETERMINING DISTANCES   Until now, we have been operating on the premise that the earth is a sphere in employing the parallels of latitude and meridians of longitude to locate positions on the earth. Now it is time to reconsider the shape of the earth and its effect on the geographic grid and the determination of distances on the earth's surface. If the earth were a true sphere and its circumference were exactly 25,000 miles (40,000 kilometers), then each of the 360 degrees would measure 69 miles, or 60 nautical miles (111 kilometers). By further subdividing each degree into minutes and each minute into seconds (and remember, we are talking about distance, not time), the results would show that 1 minute would equal 1.15 miles or 1 nautical mile (1.84 kilometer), and 1 second would represent 0.019 miles, or about 100 feet (0.03 kilometers). If we had a geographical grid position of latitude 20°30′10″ north, longitude 70°20′20″ west, we would know that the area represented would be 1415 miles (2264 kilometers) north of the equator and 4853 miles (7764 kilometers) west of Greenwich.

*For additional information on time and the rotating earth, see Appendix A.

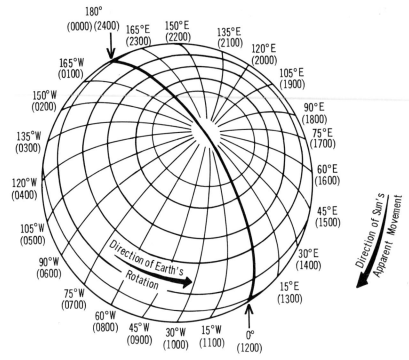

FIG. 2-6 Longitude and Time. The earth rotates about its polar axis 360 degrees every 24 hours, a 15-degree distance each hour. Time difference between the 0 meridian and the 180-degree meridian is 12 hours. When it is noon along the 0 meridian, it is midnight along the 180-degree meridian. In this illustration, the times shown relate to a noon (1200 hours) time at Greenwich, England, along the 0 (or prime) meridian.

Unfortunately, however, we cannot ignore the elliptical shape of the earth. The length of 1 degree of longitude is only 34.67 miles (55.8 kilometers). The distance of the earth (Fig. 2–7, Table 2–1). The only place where 1 degree equals 69 miles (111 kilometers) is along the equator.* Along the 60° parallel, we discover that the length of 1 degree of longitude is only 34.67 miles (55.8 kilometers). The distance between the equator and 10° north is about 687 miles (1100 kilometers), but the distance between 60° north and 70° north is only about 6 miles (9.6 kilometers) longer, thus showing us that distances measured along meridians are not greatly distorted. Consider the difference between meridians, as measured along a parallel. Distance between the prime meridian and 15° east longitude is 1035 miles (1656 kilometers) along the equator, but along the 60° north parallel, the distance shortens to 580 miles (928 kilometers)—a shortening of 455 miles (728 kilometers). If you take another look at Fig. 2–4, the reason for these variances becomes obvious: The meridians converge as they approach the poles.

* Although the precise measurement of 1 degree along the equator is 110.57 kilometers for latitude and 111.32 for longitude, geographers normally use 111 kilometers when determining distances where absolute accuracy is not required.

TABLE 2-1 Effect of Earth's shape on the geographic grid

Latitude Degrees	LENGTH OF 1 DEGREE OF LATITUDE		LENGTH OF 1 DEGREE OF LONGITUDE		Latitude Degrees	LENGTH OF 1 DEGREE OF LATITUDE		LENGTH OF 1 DEGREE OF LONGITUDE	
	Miles	Kilometers	Miles	Kilometers		Miles	Kilometers	Miles	Kilometers
* 0	68.704[a]	110.57	69.172[a]	111.32	50	69.115	111.23	44.552	71.70
↑ 5	68.710	110.58	68.911	110.90	55	69.175	111.33	39.766	64.00
↓ 10	68.725	110.60	68.129	109.64	* 60	60.230	111.42	34.674	55.80
* 15	68.751	110.64	66.830	107.55	↑ 65	69.281	111.50	29.315	47.18
20	68.786	110.70	65.026	104.65	↓ 70	69.324	111.57	23.729	38.19
25	68.829	110.77	62.729	100.95	* 75	69.360	111.63	17.960	28.90
* 30	68.879	110.85	59.956	96.49	80	69.386	111.67	12.051	19.39
↑ 35	68.935	110.94	56.725	91.29	85	69.402	111.69	6.049	9.74
↓ 40	68.993	111.03	53.063	85.40	90	69.407	111.70	0.000	0.00
* 45	69.054	111.13	48.995	78.85					

[a] A statute mile is shorter than a nautical mile by 796 feet. When traveling about the landmasses, we use statute miles, but aircraft and ships at sea use nautical miles. We are driving at 50 miles per hour, and we mean 50 statute miles per hour. An airplane traveling at 600 *knots* is traveling 600 nautical miles per hour. There are 69 statute miles for every 60 nautical miles.

SOURCE: Adapted from U.S. Geological Survey *Bulletin No. 650*, "Geographic Tables and Formulas," S. S. Gannett (1916), pp. 36–37.

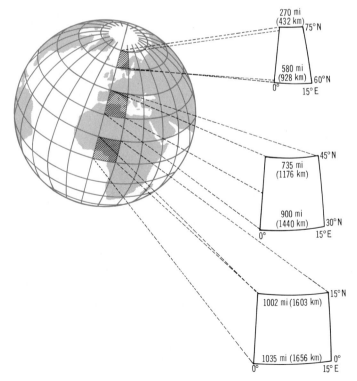

FIG. 2-7 Effect of the Earth's Shape on the Geographic Grid.

270 mi (432 km) 75°N

580 mi (928 km) 60°N
0°   15°E

735 mi (1176 km) 45°N

900 mi (1440 km) 30°N
0°   15°E

1002 mi (1603 km) 15°N

1035 mi (1656 km) 0°
0°   15°E

Maps

It was mentioned earlier that cartography is considered to be one of the earth sciences. Most people assume that cartography—the making of maps, graphs, and charts—falls directly within the province of geography. Such is not the case, however, for cartography stands on its own feet as a science—as well as being an art form requiring great skill. Cartography can be of assistance to any of the earth sciences, and it can also serve other fields: history, anthropology, botany, sociology, as well as numerous other disciplines. Yet, geographers usually are cartographers, too—not always highly skilled, but capable of evaluating maps and charts. Cartography, then, is a *tool* for geographers. Although it may be an overstatement, there is some justification in the saying, "If it cannot be explained on a map, it isn't geography." Geographers use maps to explain and show the interaction between places, people, and the various elements of nature, as well as their relationships and their distribution.

A map, however, is not the earth. It is not even that small piece of earth it purports to represent. It is *symbolic,* a representation of something too big to be observed and analyzed in a small place. A map is used to portray either the entire earth—or a small portion of it. If you were to thumb through the pages of an atlas, you might be awed by the variety of maps presented: maps showing political divisions; climatic, soil, or vegetation regions; population distribution; location of rivers; wheat-growing areas; or oceans and their currents.

Actually, one of the most important things for any of us to know is our location on this planet. If you should become lost in a strange city, you could consult a map of the city and pinpoint your position. If you plan to go hiking or camping in some out-of-the-way wilderness area, you can obtain a map of the region and find the specific locations of the places you want to visit, thus saving yourself the uncertainty of random wandering.

Today, one can go into a department store, stationery shop, or book store and discover an endless variety of maps. Perhaps it is natural for us to take this for granted. But it hasn't always been this way.

THE EVOLUTION OF MAP MAKING For thousands, perhaps millions, of years, humans knew little about the earth about them. They lived in a frightening world, eking out a precarious existence, living from day to day, wandering from place to place in the everlasting search for food. Although they had no nice little packet of maps, did they know where they were? To be sure, they had to be intimately aware of their terrain, their bit of earth space. Using the knowledge gained in their travels, they were able to fix an area's location by observing large trees, huge rocks, mountains, or streams—and maybe the stars and the sun. An individual was compelled to know where he was, else how would he dare go out on a hunt that might last for days? His only means was to observe carefully the physical features of the earth about him, so he could find his way back to his cave, where his mate and children awaited him.

The first maps probably were drawn when people gathered into tribal groups. A hunter, wishing to share with other hunters his knowledge of a valley where herds of wild game abounded, may have constructed a crude map by scratching in the dirt in front of his cave. With the passage of time, new methods evolved, such as using a piece of charcoal to mark out the location of a distant water hole on a piece of tree bark. Still later, men used fire-hardened sticks to press lines onto flat pieces of wet clay. The clay slab would then be baked over a flame, and the result was a symbolic, easily portable representation of a portion of the earth.

Archaeologists have found some of these old maps in the ruins of cities that were once the pivotal points of ancient empires, such as Arkkad, Sumer, Ur, and Lagesh in the broad valley of what was once known as Mesopotamia, the Land of the Two Rivers (Tigris and Euphrates). The rulers needed maps in order to maintain their control over the people and lands. Phoenicians, earliest of the Mediterranean peoples to sail the seas, made charts that were usable only within sight of land. The Greeks, too, made maps as they increased their knowledge and expanded their empire. Africa's Arabs, on the other hand, were not proficient in map making, although they did manage to map the positions of important oases and trading routes across the deserts.

Polynesians, who long ago colonized islands in the South Pacific, came from somewhere out of the west, perhaps from the vicinity of Southeast Asia, and sailed eastward across the mysterious ocean to the islands of the Pacific. They are not believed to have had any instruments to aid them, and their task was accomplished by skill, great daring, and luck. It was not until the Chinese invented the **magnetic compass** that ships could safely venture forth into unknown regions. While navigators had long used the North Star and the sun to determine their positions at sea, imagine the difficulties they faced when the sky was heavily overcast or when storm clouds obscured the rocky headlands and dangerous shoals. The introduction of the compass made it possible for seafaring men to travel anywhere, any time. They could sail beyond the horizon and discover new worlds to conquer, to exploit, to settle. The compass and the map: indispensible tools for mankind.

When the Romans began to ride the seas, they became excellent mapmakers, and it is interesting to note, our term *map* comes to us from the Latin, *mappa,* which means "cloth" or "napkin." Is it possible that Roman seamen drew sketches of their travels on the tablecloths of their favorite seafront grog shops? Were those tablecloths later used as maps?

ROUND GLOBE TO FLAT MAP The only valid representation of the earth is the globe. It is spherical, as is the earth, and it can help you to gain an understanding of the earth's shape, the location of the various continents, and the geographic grid, as well as to visualize the reasons behind the changing of the seasons. However, the globe has one great drawback: It is difficult to cart around. Can you imagine taking a globe with you on a round-the-world flight or trying to fit it into the glove-compartment of your car? This is, of course, the reason we use flat maps. Maps may not be as accurate as a globe with regard to distance and direction or in the size of areas in different parts of the world, but they are convenient to insert into textbooks, newspapers, or one's pocket. Their drawbacks are, especially, in their inherent distortions. David Greenhood, in his book, *Mapping,* explains it this way: "Here, then, is an instance where the unscrupulous propagandist can get in his dirty work. He can give a picture of one country being impressive or insignificant as compared to another by using a map in which the countries are not shown in their truly equivalent sizes."[5] And this warns us that if we are really to understand a map, we must know what properties it must have.

How can you flatten a round object without changing its characteristics? To understand the problem, let's pretend an orange, the kind with a thick peel, is a globe. With a marking pen, draw the parallels, meridians, and continents on the orange peel. Then cut down each line of meridian almost to the equator. Carefully peel the orange (Fig. 2–8) and flatten out the peeling. What does it look like? Does it still resemble the globe? What about the large empty spaces?

This is the problem that confronted mapmakers after they came to believe the earth was round: How can we turn a round globe into a flat map and not have a lot of voids? If the meridians are supposed to run in a true north-south direction and converge at the poles, and the parallels are supposed to run in a true east-west direction, be equidistant, and remain parallel to each other and the equator, how does the mapmaker keep them that way on a flat map? No one has as yet found the perfect answer, because all flat maps must have some distortion (Fig. 2–9). From this, we can infer that the maps with which we are acquainted have been manipulated, altered, and modified in order to be of use to us. This means that cartographers must consider which of the qualities a globe possesses are most important for the job at hand. He must first ask himself which of the various properties must be emphasized: Should the areas on the map be equal to each other in area, just as they are on a globe? But what if the areas will not look just as they do on a globe? Should the map be constructed so that an airline navigator can take true direction from his map? These questions must be answered before the cartographer can proceed to construct his map projection.

FIG. 2–8 Round Globe to Flat Map: Peeling an Orange.

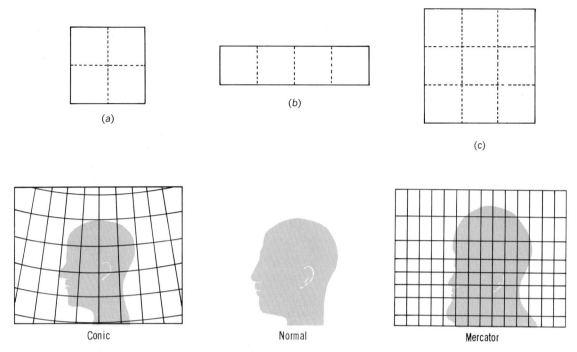

(a)

(b)

(c)

Conic   Normal   Mercator

FIG. 2-9 Flat Maps Cannot Duplicate the Globe.   Some maps are conformal, which means that areas on the map have the same *shape* as they do on the earth or the globe. Look at (*a*) and (*c*). (*a*) and (*c*) are not the same size. They differ in the amount of *area* enclosed. (*a*) and (*b*), however, enclose the same amount of area (earth space) even if they do not have the same shape; thus, (*a*) and (*b*) are considered equal-area. All maps are distorted in some way, so one uses a particular projection that is designed for a particular purpose. For example, the Mercator projection is used for ship navigation; azimuthal projections, for air travel; and conic projections, for maps that portray midlatitude regions such as the United States or Europe.

PROPERTIES OF MAP PROJECTIONS   Properties of map projections include the following: equal-area (equivalence), conformality (conformal), equal-distance (equidistant), and azimuthal (zenithal). Some map projections are limited to just one of these properties, some have two or more, but cartographers try to soften the deviations from a global representation. If the shapes of islands or coastlines are to be shown "correctly" (to look as they do on a globe), the property to be attained is **conformality**. If areas conform to their actual shape, South America and Greenland will look as they do on a globe. They will not be shown in the proportion to other areas or to each other—as you can see by examining a conformal projection, such as the one developed by Mercator in the sixteenth century (Fig. 2–10). Compare these two landmasses as to shape and size, then examine them on a globe. Is Greenland really larger than South America? It might be noted that there is less distortion of small areas on the map and areas along or near the equator.

Map projections that are considered **equal-area** are those portraying areas in the

35

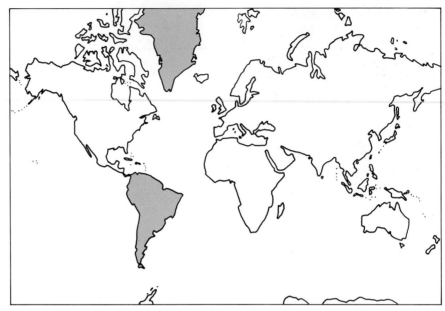

FIG. 2-10   Mercator's Projection: Conformality Emphasized.

same relation to each other as they are on a globe. This means that all areas on the map are in the proper proportion to each other. On an equal-area world map, 1 square inch measured over North America will enclose the same amount of area as 1 square inch measured over Africa or Siberia. In other words, areas on the map will enclose the same amount of space they would on a globe (Fig. 2–11).

If accuracy in measuring distance is important for the projection, the property of **equal-distance** is emphasized, which means that distances can be measured with

FIG. 2-11   Mollweide's Projection: Equal Area Emphasized.

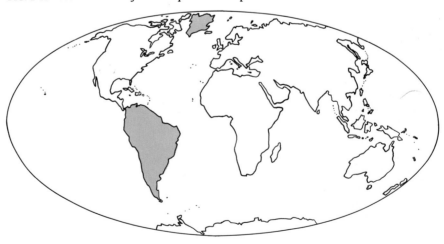

accuracy from the focal point of the map (Fig. 2–12). This focal (or central) point can be set at one of the poles, or it can be placed over any location on the earth. Many aerial navigation charts (maps) are constructed with this emphasis. Some, however, are more concerned with direction, and the property of distance will be sacrificed. In this instance, the projection is **azimuthal,** and all directions taken from the center are true compass directions. If the center were placed on New York's Kennedy International Airport, the navigator could lay out his flight path with ease.

A map, then, is a concession: It is not the whole truth. Each of the map projections we shall consider later in this chapter have been formulated to handle a specific job—to be used for ship or air navigation, to portray the entire earth, to focus on certain midlatitude regions, to be used in newspapers, or to be used in textbooks. A projection engineered to portray the United States does not have to be the same kind needed to portray the world.

It might be well for us to examine some of the more important elements in the construction of map projections.

MAP ESSENTIALS   Once a cartographer has settled on the kind of projection he is going to use, he then asks a series of questions: How much of an area am I going to show? A city or a portion of a city? A state or a larger region encompassing several states? An entire continent—or a hemisphere? Or the whole earth? What scale shall I use—large or small? How shall I indicate directions? What symbols are needed? What will be my title? Figure 2–13 indicates one cartographer's answer. Let us consider these essentials.

Map scale   Maps are not merely pictures drawn to a person's whim; they are representations of the earth or a portion of it. They must not only be realistic, but

FIG. 2-12   Azimuthal Map Projections.   (a) is a zenithal projection with the central, focal point at the geographic North Pole; (b) has its focal point at Tokyo, Japan. Both are equal-distance map projections.

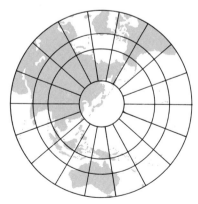

(a)   (b)

they must also be accurate in as many respects as possible. One way the cartographer attempts to make his map meaningful is to employ a scale indicating the ratio between distances on the map and distances on the earth space being represented. There are three types of map scales in use today; some maps indicate only one, some have all three (Table 2–2). Any of these scales can be used for very small areas (such

FIG. 2-13 Map Essentials.

URBAN AREAS

⊛ over 6,000,000 (national capital)
● over 3,000,000
◉ over 1,000,000
• 500,000 - 1,000,000
○ 250,000 - 500,000 (selected places)
· selected smaller places

POLYCONIC PROJECTION

THE SOVIET UNION

0   500   1,000 miles

Legend   Type of Projection   Title   Scale   Latitude and Longitude

38

as your college campus) or for extremely vast areas (such as the Western Hemisphere or the entire earth).

Figure 2–14 illustrates the difference between large- and small-scale maps. If the size of the area to be portrayed on a map is small, and a considerable amount of detail must be shown (such as houses, factories, roads, paths, farms, railroads, power-lines, churches, schools, and so forth), a **large scale** will be used. As an example, on some large-scale maps, 1 inch (2.5 centimeters) equals 24,000 inches (60,000 centimeters), or 0.26 miles (0.42 kilometers). If, however, a much larger portion of the earth's surface is to be portrayed (such as a province or state, nation, hemi-sphere, or the entire world), the amount of detail will be severely reduced, since the cartographer must compress a vast area onto one small page. On some **small-scale** maps, 1 inch (2.5 centimeters) equals 100,000,000 inches (250,000,000 centimeters), or 1600 miles (2560 kilometers).

Although the verbal and graphic scales are self-explanatory, the representative fraction (R.F.) scale is sometimes confusing. To clarify it: The bottom number (the denominator) holds the key. The larger the denominator, the smaller is the scale. An R.F. scale of 1/15,000 is a much larger scale than one calling for 1/15,000,000. To further clarify, we can say that on a map of 1/15,000, 1 inch would equal 15,000 inches (about 0.23 miles or 0.36 kilometers). On a map with an R.F. scale of 1/15,000,000, 1 inch would equal 15 million inches or about 236 miles (377 kilometers). Obviously, the R.F. with the smaller denominator is a much larger scale. Just as with any fraction, the smaller the denominator, the larger the fraction: Thus, $\frac{1}{10}$ is larger than 1/10,000,000.

Maps legends and symbols   Any map that has something to say, some information to get across, has a title. For example, thumb through the pages of this book and

TABLE 2-2   Map scales

A map scale is the ratio of the distance on a map as it corresponds to the actual distance on the earth's surface. Without an accurate scale, a map is useless and unreadable.

VERBAL SCALE.   Also known as the *stated scale*. The ratio is given in words:
"1 inch (2.5 centimeters) equals 100 miles (160 kilometers)," or, "1 inch (2.5 centimeters) on the map represents, or is equal to, 100 miles (160 kilometers) on the earth's surface."

GRAPHIC SCALE.   Also known as the *bar, line,* or *linear scale*. The ratio is shown as a line marked off in units (specified intervals or increments):

REPRESENTATIVE FRACTION SCALE.   Also known as the *R.F. scale*. The ratio is given as a fraction without specifying a definite unit (such as inches, centimeters, yards, meters, miles, or kilometers). The ratio states (1:63,360 or 1/63360) that one unit on the map represents a specified number of the same kind of unit on the earth's surface. For example, 1:63,360 could mean that 1 inch on the map equals 63,360 inches on the earth.

FIG. 2-14 Large-Scale to Small-Scale Maps.

Small scale 1 in.= 1600 mi

Medium scale 1 in.= 160 mi

Large scale 1 in.= 16 mi

examine the various maps. Each has a title, a caption that tells you what the particular map is attempting to explain. A title alone, however, usually is not sufficient. There must be additional explanation of just what all the symbols—colors, dots, cross-hatching, or shadings—mean. This is the job of the map's **legend** (Fig. 2–15). The purpose of the legend is to explain those symbols. Figure 2–15(*c*) uses dots to show the distribution of human population on the face of the earth. Figure 2–15(*a*) uses graduated circles to show the relationship of one area or place to another, regarding percentages or the proportion of a segment of the population of a city to the total. Figure 2–15(*b*) uses colors and numbers to explain the distribution of soils.

On many of the maps using color, certain colors seem to be standardized. For example, blue usually denotes water features, such as streams, rivers, lakes, seas, or oceans. On certain maps (such as the topographic maps made by the U.S. Geological Survey), green is used for vegetation, black is used for buildings and cities, and red double lines are used for major highways. Some maps have special symbols, such

as ⛪ for churches, ⚑ for schools, 〜 for intermittent streams. For additional examples, see Fig. B–2 in Appendix B.

Later in this text you will find another type of symbolism: lines sweeping across the map, swerving this way and that—boundary lines that connect points of equal value and separate areas of different values. Some geographers call these lines **isolines**; others refer to them as *isopleths* or *isarithms*. The root *iso* comes to us from the Greek *isos,* meaning "equal." Isarithms refer to lines having an exact value base, and isopleths are those having a relative or nonabsolute value. **Isotherms** (see Figs. 4–4 and 4–5) are lines on a map connecting points that have (at a specific point in time) the same air temperature. **Isobars** (see Figs. 4–7 and 4–8) are lines connecting places with equal atmospheric (barometric) pressure. These various isolines will be discussed frequently in later chapters.

Map direction   When you pick up a map, do you immediately know where "north" is? Do you realize that there is not just one north, but three? Is north one particular spot on the earth? If so, how could you pinpoint north on a projection such as is shown in Fig. 2–16(*a*)? If every meridian runs in a true north-south direction, how can there be an infinite number of norths? Obviously, there cannot be. If you check Fig. 2–16(*b*), the Northern Hemisphere, you can see "true north," which is more properly called the **geographic north**. Now, what happens when you hold a magnetic compass in your hand as in Fig. 2–16(*c*)? Does the magnetized needle point toward the geographic north? No, because it aims toward a large underground deposit of iron located on Prince Edward Island, Canada (latitude 73° north, longitude 99° west). The difference (which we shall cover in more detail in Appendix B) is called the **magnetic variation** (or *declination*). This variation does not remain constant, because the magnetic north is always changing slightly. We must make allowances for the variation, and many maps indicate the number of degrees your compass needle will diverge from geographic north. If you were to crisscross North America, checking carefully to determine the exact amounts your needle was "off," you would come up with a map (Fig. 2–17) covered with **isogons** (lines connecting

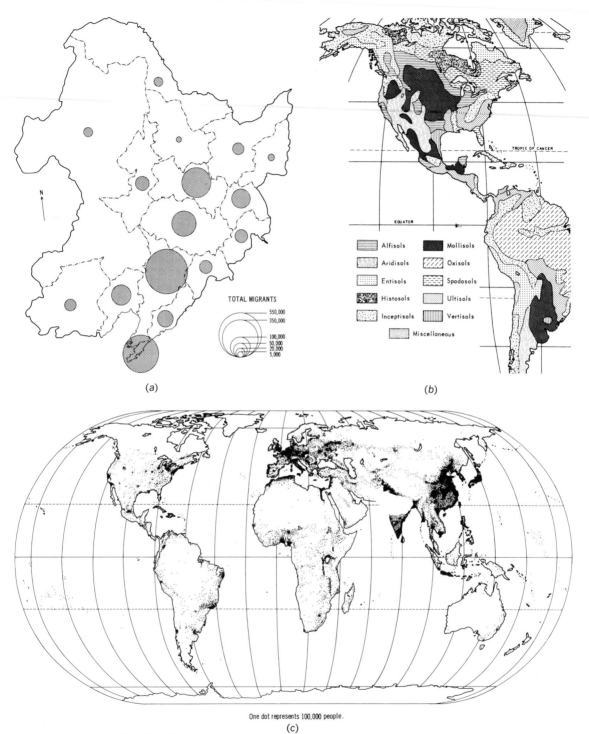

TOTAL MIGRANTS

550,000
350,000

100,000
50,000
20,000
5,000

(a)

TROPIC OF CANCER

EQUATOR

Alfisols	Mollisols
Aridisols	Oxisols
Entisols	Spodosols
Histosols	Ultisols
Inceptisols	Vertisols
Miscellaneous	

(b)

One dot represents 100,000 people.

(c)

FIG. 2–15 Map Legends and Symbols.

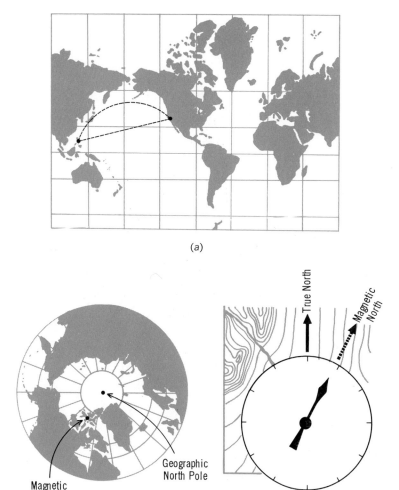

FIG. 2-16 Map Directions.

points of equal magnetic variation away from the geographic north). You would then be able accurately to use your magnetic compass with your maps. You will note that there is one isogon marked 0°—where there is no magnetic variation (or declination from the geographic north). This is called the **agonic line,** and it is the only isogon where all points on it line up with both the magnetic and the geographic North Pole.

The last of the norths is called **grid north,** and this refers to the "bending" of the meridians while laying out the United States Land Survey System (see Appendix C, "Ground Referencing Systems"). Ownership of parcels of land was determined by establishing an initial, or principal, meridian in the area being surveyed (based on geographic north) and a baseline (a parallel of latitude). The lines bordering properties

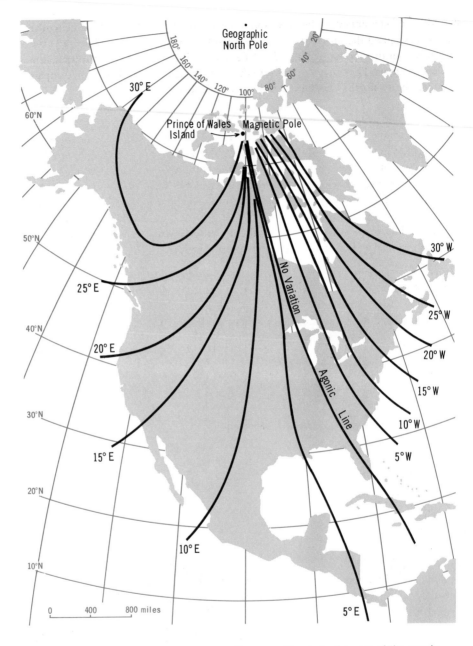

FIG. 2-17 Isogonic Map of North America. If your position is to the west of the agonic line, your compass will point to the right (to the east) of true north. It will point toward the magnetic North Pole.

should have been aimed at the geographic north, but the local need for rectangular coordinates caused the surveyors to ignore (slightly, maybe as much as 3 degrees) the true north-south meridianal lines. You might say this is a warning not to trust the borders of a map: They may not point to the geographic north. If there is no compass rose to indicate north, you should follow the meridians.

TYPES OF MAP PROJECTIONS   Our purpose here is not to compete with the many excellent texts on cartography,* but merely to acquaint you with some of the ways cartographers have tried to portray our earth. There are two basic ways to develop a map projection, and in both methods the same result is sought: to have the parallels and meridians intersect exactly as they do on a globe. One way is to project the round face of the earth onto a flat plane by using a lighted (from the inside) globe. The parallels, meridians, continents, islands, and oceans are cast like shadows onto the plane. The other way is to use mathematics (geometry and trigonometry) to produce a special projection suited for a special purpose.

Cylindrical projections   To construct **cylindrical projections,** an internally lighted, transparent globe is placed inside a cylinder composed of something like clear plastic (Fig. 2–18). When the light is turned on, the lines representing the parallels and meridians, as well as the continents and other landforms that have been printed on

* Those interested might consult: David Greenhood (1964), *Mapping* (Chicago, Ill.: University of Chicago Press); Erwin Raisz (1948), *General Cartography* (New York: McGraw-Hill); A. H. Robinson and R. D. Sale (1969), *Elements of Cartography,* 3d ed. (New York: Wiley); Judith Tyner (1973), *The World of Maps and Mapping* (New York: McGraw-Hill).

FIG. 2–18   Construction of a Cylindrical Projection.

Line of Tangency

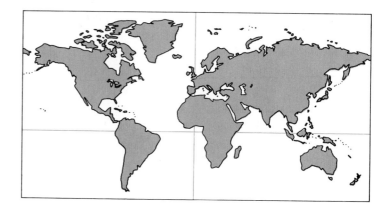

the globe, are projected onto the covering cylinder (just as movies are projected onto a screen). The line where the cylinder touches the globe is called the **line of tangency,** and only along this line (or a few degrees on either side) is there accuracy. Away from the line of tangency, the parallels are spaced further and further apart, and different scales must be used for measuring distance. As for the meridians, they no longer converge as they approach the poles. Instead, they are parallel to each other and are equally spaced from the equator to the poles.

Of all the cylindrical projections constructed, only the Mercator is conformal, and it is used primarily for ship navigation. This is because navigators can ignore the tremendous distortions, since a series of straight lines drawn along the route from one port to another will cut every meridian at the same angle. Thus, the ship has only to steer a compass course along that line (with only slight direction changes), and it will travel the shortest distance between the two ports.*

Conical projections   To construct **conical projections,** a clear plastic cone is placed over an internally lighted globe (Fig. 2–19) with the apex of the cone placed atop the globe. For a simple conic projection, the cone touches the globe along one line of tangency (standard parallel). When the light is turned on, the geographic grid and landforms printed on the globe are transferred to the plastic cone. The meridians appear just as they do on the globe—they converge toward the pole. Parallels, however, are not equally spaced, unless the cone intersects the globe at more than one parallel, in which case a *modified conic* projection results, one that has less distortion of areas.

Another form of conical projection was developed by a cartographer named Lambert in the eighteenth century, the Lambert conic conformal (Fig. 2–20).

*The line that intersects all meridians at the same angle is called a *rhumb line,* or *loxodrome.*

FIG. 2-19   Construction of a Conical Projection.

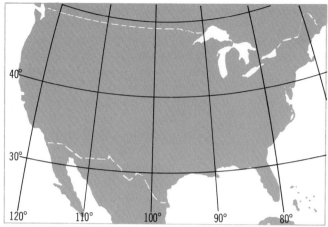

FIG. 2-20 **Construction of Lambert's Conformal Conic Projection.**

Lambert laid out the parallels mathematically, so that the shapes of the continents were conformal. This projection is widely used to portray midlatitude regions, such as the United States, Europe, or the Soviet Union.

Azimuthal projections For **azimuthal projections,** a flat transparent sheet is placed over a lighted globe (Fig. 2–21), and here, the point of tangency is the center of the projection. In the case of a polar-azimuthal projection, the meridians radiate out

FIG. 2-21 **Construction of an Azimuthal Projection.**

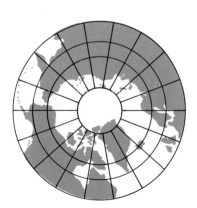

from the central point (the pole) like spokes on a bicycle wheel. Lines drawn from that point to any other location on the map are straight lines, part of a great circle. There is also a true compass direction from that point. Azimuthal projections can have any point on earth for their central, or focal, point, and then we use different names: *oblique, azimuthal equal-area, orthographic, polar stereographic, equatorial stereographic,* and so on. All of these designations are useful, and each projection has a specific use—whether they portray entire hemispheres and continents or relatively small areas, and whether they are for newspaper reproduction, airline navigation charts, or ocean sailing charts.

Miscellaneous projections Included in the category of miscellaneous projections are maps that have a combination of properties, and that have been formulated through mathematical equations. In some cases, the meridians and parallels have been manipulated to suit the mapmaker's purpose. A good example of this would be the homolographic projection, in which the meridians have been turned into ellipses, and the globe has been stretched out from side to side. The distance between the poles is about half the distance along the equator, yet the projection remains equal-area. One of the most popular of these projections among today's geographers was developed by Dr. Paul Goode. It is an interrupted homolosine (Fig. 2–22) projection, which means that it is a combination of two different mathematical projections, homolographic and sinusoidal. In this projection, each continent is centered on its own meridian, and the distortion of the landmasses is minimized in the higher latitudes by splitting the oceans.

"Eckert IV" (Fig. 2–23), the map projection for this text, is an equal-area

FIG. 2-22 Goode's Interrupted Homolosine Projection. (Copyright by The University of Chicago, Dept. of Geography.)

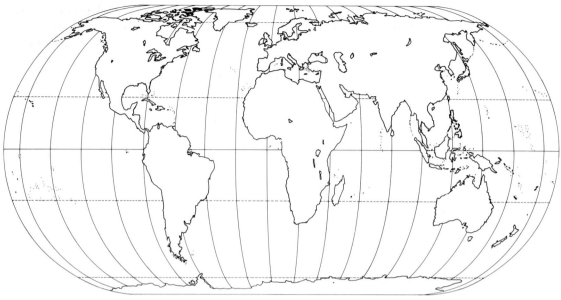

FIG. 2-23 Eckert IV Projection.

projection devised by Max Eckert to minimize distortion and crowding of the landmasses and oceans in the higher latitudes. Despite some shearing of the continents and islands near the polar regions, this projection is an excellent one for showing world distributions. The parallels are straight lines that are mathematically designed to diminish in their spacing toward the poles. The meridians are elliptical, and they do not converge to a point at the poles; therefore, the poles are represented as straight lines that are one-half the length of the equator.

Cartograms Yet another category of map projections is cartograms, where distortion is deliberate. In this type of projection, no attempt is made to make the maps conformal, equal-area, equidistant, or azimuthal. Some are intended to be humorous. For example, think what the results would be if someone asked a resident of Brooklyn—someone who had never left his own neighborhood—to make a map of the United States. The result shows the influence of movies, television, and magazines on the mind of an untraveled person.

Sometimes a map is made to illustrate some special point or feature, such as the population distribution of the world, country-by-country. A map like this would be quite unlike the representation in Fig. 1–1. There would be no dots or circles to show population density, for each country is drawn to a size that is in proportion to its population. Meridians and parallels in this projection are figments of the artist's imagination, added to make the cartogram more "realistic." It is an interesting way to point out the vast disparities between land area and population.

THE City of NEW YORK is unique – it is a nation within a NATION. Its inhabitants, of which there are some 7,000,000, are called NEW YORKERS. This MAP is presented, after patient research, as a composite of the NEW YORKERS' ideas concerning THE UNITED STATES ··

LET THEM SPEAK

We have cousins in the West·· They live in Wilmington, Delaware.

He is moving to Dallas so he can be near his little Mother in El Paso

Indiana was an Indian Reservation until just recently, wasn't it?

So you are moving to Indianapolis; you must let me give you a letter to my niece in Minneapolis.

Oh yes! he entered the Marathon Swim from Los Angeles to Hawaii · · · · ·

A New York
THE UNI
OF A

Copyright by Daniel K. Walling

NEW YORK WORLD'S FAIR
1939

ker's Idea of
ED STATES
ERICA

452 West 144th Street, New York, N.Y.

Physiographic diagram Another way to portray the earth in a graphic way is through physiographic diagrams. Here, the cartographer attempts to picture the land as it might look from space. Pioneers in this field were A. K. Lobeck and Erwin Raisz.*

With improvements in cameras in outer space, it may not be long before our maps will be actual photographs of the earth's surface—with the cultural features superimposed for clarification. We have only begun to use these magnificent photographs, and already geologists have discovered fracture zones and surface conditions never before suspected. In addition, the new field of remote sensing (which we shall discuss in Appendix B) offers tremendous possibilities for mapping, as well as for exploration of our planet. The future of aerial mapping promises to be exciting.

Review and Discussion

1. Explain the term *line of sight.*

2. After doing some additional research on the experiments conducted by Eratosthenes, explain his method for determining the circumference of the earth.

3. Why do you suppose it took so long for humans to accept the theory that the earth was a sphere?

4. What error did Ptolemy commit in making his map projection that caused Columbus to reach the New World rather than China?

5. Who was the first man of record to actually construct a globe?

6. Explain these terms: *sphere, spheroid, ellipse, ellipsoid, oblate, oblate ellipsoid, geoid.* Which do you prefer?

7. Briefly define the term *geographic grid.*

8. Explain the difference between great circles and small circles.

9. What is a meridian? Explain the term *meridian of longitude.*

10. What is the prime meridian?

11. What is a parallel of latitude?

12. How many parallels are great circles? How many meridians are great circles?

13. What is a sextant? How does it aid in determining the latitude of a place?

14. How can you use the North Star (Polaris) to determine your latitude?

15. What do meridians of longitude have to do with time?

16. What times of day are the following: 1800; 2300; 0800; 1200; 2400?

*Among some of the fine works produced by Erwin Raisz, see his (1948), *General Cartography* (New York: McGraw-Hill); (1956), *Mapping the World* (New York: Abelard-Schuman); and (1962), *Principles of Cartography* (New York: McGraw-Hill).

17. What effect does the earth's ellipsoidal shape have on the distance between meridians of longitude?

18. Discuss the evolution of map making.

19. Why is the magnetic compass important?

20. Why is an understanding of map distortions important?

21. Explain these map projection properties: equal-area, conformality, equal-distance.

22. What is meant by the term *azimuthal*?

23. Explain the difference between large-scale and small-scale maps.

24. Explain these map scales: verbal, graphic, representative fraction.

25. Why is the map legend important?

26. What is an isoline?

27. What is the relationship of the isogonic lines to magnetic variation?

28. What is the difference between grid north and geographic north?

29. What are the main differences between cylindrical and conical projections?

30. What is a rhumb line?

Notes

[1] Robert O. Ballou, ed. (1967), *The Portable World Bible* (New York: Viking), p. 53.

[2] Jacquetta Hawkes (1962), *Man and the Sun* (New York: Random House), p. 61. © Copyright, 1962, by Jacquetta Hawkes. With permission.

[3] *Larousse Encyclopedia of Mythology* (1959), trans. Richard Aldington and Delano Ames (London: Paul Hamlyn), p. 11. Reproduced by permission of the Hamlyn Publishing Group Limited from *Larousse Encyclopedia of Mythology*.

[4] Robert C. Toth (25 July 1969), "Flat Earthist Leader Refuses to Give Up" (UPI), *Los Angeles Times.*

[5] David Greenhood (1964), *Mapping* (Chicago, Ill.: University of Chicago Press), p. 117.

Additional Reading

Brown, Lloyd A. (1949). *The Story of Maps.* Boston, Mass.: Little, Brown.

Bunbury, E. H. (1959). *A History of Ancient Geography.* Vols. 1 and 2. New York: Dover Publications.

Raisz, Erwin (1962). *Principles of Cartography.* New York: McGraw-Hill.

Robinson, A. H., and R. D. Sale (1969). *Elements of Cartography.* 3d ed. New York: Wiley.

Thrower, Norman J. W. (1972). *Maps and Man. An Examination of Cartography in Relation to Culture and Civilization.* Englewood Cliffs, N.J.: Prentice-Hall.

3

Spaceship Earth

The waters deluge man with rain, oppress him with hail, and drown him with inundations; the air rushes in storms, prepares the tempest, or lights up the volcano; but the earth, gentle and indulgent, ever subservient to the wants of man, spreads his walks with flowers, and his table with plenty; returns, with interest, every good committed to her care; and though she produces the poison, she still supplies the antidote; though constantly teased more to furnish the luxuries of man than his necessities, yet even to the last she continues her kind indulgence, and, when life is over, she piously covers his remains in her bosom.

PLINY THE ELDER

Many of the ancient peoples thought of the earth as a "supermother," who responded to the loving attentions of a "sky father"; these two, of course, looked after human needs. Later civilizations thought of the earth as a motionless body, hanging suspended in space, surrounded by the moving stars, planets, moon, and sun, all of which made regular orbits around the earth.

Today, however, we think of the planet earth more as a spaceship hurtling through the vast, unknown reaches of the universe. Astronomers and earth scientists realize that every body in space is moving, and each has its own path. The earth is to the universe as a drop of water is to the world ocean.

Our task now is to examine the role the earth plays in the immense scheme of things, and the reasons behind earth's movements. Some of the questions we shall attempt to answer apply directly to the earth's power plant, the engine that makes it function so smoothly—the sun. We shall examine the relationship between the sun and the earth, the controls over the heating and cooling processes, and the effect of the changing seasons on earth and its life forms.

Earth and the Solar System

Despite the wishful thinking of the ancient philosophers, our earth is not the center of the universe. It is but one tiny part of one solar system. With the sun at the center, planets extend billions of miles into space (Fig. 3–1). Diminutive Mercury is the closest to the sun, followed in order by Venus, Earth, Mars, Jupiter, Saturn, Uranus, Neptune, and Pluto. Our present position and size in relation to the sun, and our present stage in planet evolution, provide us with surface and atmospheric conditions that allow plant and animal life, in the forms we know, to survive and evolve. It is quite possible, as some scientists suggest, that there are millions of "earths" like ours revolving around stars or suns somewhere out there in space, but whether they have life forms with which we are familiar is a debatable question. If our space program continues, perhaps someday we shall know.

Our position in the solar system is a delicate one. We are, evidently, in exactly the right place—neither too close nor too far from the sun. Mercury, on the other hand, is much too close. The *Mariner* flight passed this small planet in March 1974 and sent us photographs of what appears to be a dead and sterile planet, barren of what we might consider life forms. It is about one-third of earth's distance from the sun, and daytime temperatures soar into the hundreds of degrees centigrade; at night the temperature falls to hundreds of degrees below freezing. Almost the

FIG. 3–1 Our Solar System. Distances of the planets from the sun are given in millions of miles. (Distances in kilometers are given in parentheses.) (Not drawn to scale.)

Pluto
3675 m. mi
(5880 m. km)

Neptune
2797 m. mi
(4475 m. km)

Uranus
1787 m. mi
(2859 m. km)

Saturn
887 m. mi
(1419 m. km)

Jupiter
484 m. mi
(774 m. km)

Mars
142 m. mi
(227 m. km)

Venus
68 m. mi
(109 m. km)

Earth
93 m. mi
(149 m. km)

Mercury
36 m. mi
(58 m. km)

Sun

The Planet Mercury. This view of Mercury is a photo-mosaic that was constructed of 18 photos taken at 42-second intervals by *Mariner 10* on March 29, 1974.

The Red Planet, Mars. Three photos of the Northern Hemisphere make up this photomosaic taken by the *Mariner 9* spacecraft in August 1972.

same situation exists on the planet Venus. Although this planet is about the same size as the earth, it apparently makes only one rotation on its axis each year, and its surface temperature was measured by *Explorer II* at over 600°F (316°C). We know little of Venus's surface, because of its dense cloud cover.

Mars, however, has surface temperatures somewhat akin to those of earth. We have assumed Mars to be a lifeless planet, yet our space probes have indicated there is some atmosphere present. Only time and continued exploration will reveal whether there is enough oxygen and nitrogen in the atmosphere to support life, or whether it is mostly carbon dioxide. As for other planets, we do not, at present, have sufficient information to project answers.

To understand why our earth has such a favored position over the other planets, we look to the science of astronomy. The astronomers tells us that the source of all our energy is the sun. The sun is a spherical body composed of boiling gas and leaping flames, with surface temperatures reaching 10,000°F (5537°C). It operates like a gigantic furnace, furnishing the earth with the energy that makes life possible. One might ask: How does the earth remain in its position, neither falling into the sun nor flying away from it out into space? For the answers, we shall take a closer look at the movements of the earth: its orbit around the sun (revolution) and its spinning around its axis (rotation).

EARTH'S REVOLUTION   The earth revolves (orbits) around the sun in an elliptical (oval) path, which causes it to swing outward to its furthest distance from the sun **(aphelion)** on or about July 4, then to close in on the sun to its closest approach **(perihelion)** on or about January 3.* Figure 3–2 portrays this great looping path. The earth's orbital speed varies (although its rotational speed around its axis remains constant), and the earth slows down as it approaches aphelion and speeds up as it nears perihelion. With respect to the sun, the earth takes $365\frac{1}{4}$ days to complete a full orbit. This is called the **tropical** (or solar) **year** and requires us to add a day once every four years (leap year) on February 29 to make up for our use of 365 twenty-four-hour days. If we were to calculate the duration of the orbit by relating to a distant star, the orbit would take only 365 days. This is called the **sidereal year**— for the term *sidereal* means "with respect to the stars."

Astronomers are very precise. They make periodic checks of the earth's actual position sidereally, and on New Year's Eve 1972, they delayed the start of the new year by one second in order to bring earth time into proper alignment. The sidereal day is exactly 23 hours, 56 minutes, 45 seconds long. The 24–hour day to which we are accustomed came to us from ancient Sumeria in the Middle East. The Sumerians used the numbers 12 and 60 as their basic measurement, allowing 12 hours for day and 12 hours for night. The idea of *12* came from their observation of lunar months.

If our earth were to swing too far away, we would not receive the necessary amount of solar radiation to sustain our forms of life. The earth might slowly freeze, and all things would die. On the other hand, if we came too close to the sun, our atmosphere and our waters would boil away. The earth would become a lifeless, inert mass, floating through space to become eventually, perhaps, similar to the planet Mercury. How, then, does the earth manage to avoid these catastrophes and remain just the right distance from the sun? The answer to that lies in two opposing, counteracting, balancing forces called *inertia* and *gravity.*

As everything is a magnet, masses of matter are attracted to one another (a principle called the *universal law of gravitation*). The power and the strength of the magnetic force (gravity) depends on the *mass* (size and weight or bulk) of the object. A smaller or lighter mass will tend to be pulled toward—or gravitate to—a larger or heavier mass (and vice versa, to a lesser degree). Thus it is with the moon in its relation to the earth, and the earth in its relation to the sun. If it were not for the balancing force called inertia, the moon would crash onto the earth, and the earth would swing down into the fiery maw of the sun.

*The terms *aphelion* and *perihelion* come to us from the terms *apogee,* the point at which an orbiting body is at its furthest distance from the body being orbited, and *perigee,* the point of closest approach. *Helion* refers to the sun and comes from the Greek word *helios*—the sun.

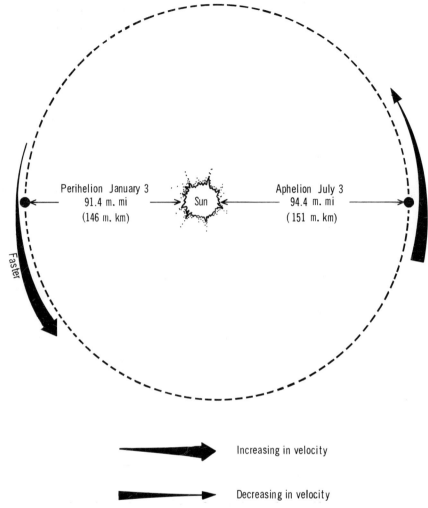

Perihelion January 3
91.4 m. mi
(146 m. km)

Sun

Aphelion July 3
94.4 m. mi
(151 m. km)

Faster

Increasing in velocity

Decreasing in velocity

FIG. 3-2 Earth's Orbit around the Sun. The shape of the arrows indicates both direction and velocity of the earth in its orbit.

To explain this force, in 1687 the English mathematician Isaac Newton formulated his first law of motion. In this law, Newton explained that a moving body will try to resist any outside or external force that might try to move it from its path (Fig. 3-3). This means that the earth is trying to move through space in a straight line, but the sun is exerting a gravitational pull that the earth cannot resist. The earth, therefore, moves in a looping, elliptical path around the sun. It is a constant, never-ending struggle—one force against the other—and fortunately for us, neither force so far has gained the upper hand. The same struggle occurs here on earth. Inertia tries to fling humans, animals, buildings, and even our atmosphere, off and away

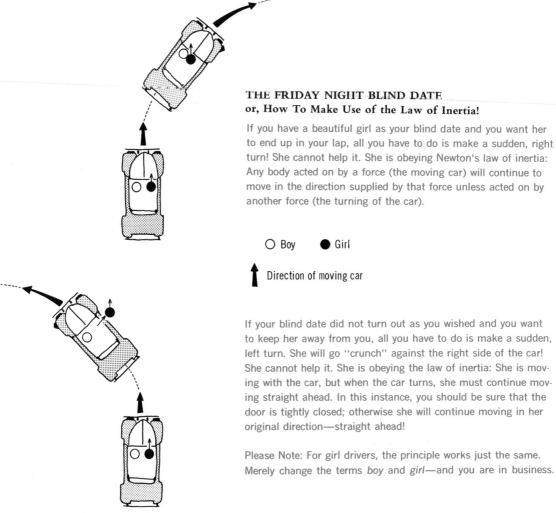

THE FRIDAY NIGHT BLIND DATE
or, How To Make Use of the Law of Inertia!

If you have a beautiful girl as your blind date and you want her to end up in your lap, all you have to do is make a sudden, right turn! She cannot help it. She is obeying Newton's law of inertia: Any body acted on by a force (the moving car) will continue to move in the direction supplied by that force unless acted on by another force (the turning of the car).

○ Boy   ● Girl

⬆ Direction of moving car

If your blind date did not turn out as you wished and you want to keep her away from you, all you have to do is make a sudden, left turn. She will go "crunch" against the right side of the car! She cannot help it. She is obeying the law of inertia: She is moving with the car, but when the car turns, she must continue moving straight ahead. In this instance, you should be sure that the door is tightly closed; otherwise she will continue moving in her original direction—straight ahead!

Please Note: For girl drivers, the principle works just the same. Merely change the terms *boy* and *girl*—and you are in business.

FIG. 3-3   Newton's First Law of Motion: Inertia.

from the earth's surface and out into space, but the gravitational attraction of the earth overcomes that force—and here we stay.

EARTH'S ROTATION   While the earth is revolving around the sun in its orbit, it is at the same time moving in another way: It is rotating (spinning) eastwardly around its axis (an imaginary line that pierces the earth through the poles, north to south). The speed of rotation at the equator is about 1022 miles (1635 kilometers) per hour, slowing down to zero at the poles. This is very fast, but many scientists believe that our present rotational speed is considerably slower than it was several billion years ago. They say that it is highly probable that the earth rotated around its axis once every five hours. Imagine a $2\frac{1}{2}$-hour day, followed by a $2\frac{1}{2}$-hour night. Think of what the earth must have looked like then: a really oblate spheroid—much wider at the equator than it is now.

Do we feel this speed? No, because everything about us—the air, clouds, trees—firmly held to the earth by gravity, is moving *with* the earth. Yet, we know that there is movement, even if we cannot feel it, for it causes night and day. The earth rotates around its axis once every 24 hours, and this continuous rhythm has a tremendous influence, not only on life conditions, but also on the direction and speed of the winds and the oceans, the length of the day, and the apparent sunrise and sunset. The relative uniformity of time during which areas of the earth receive solar radiation might be likened to the even cooking of a chicken revolving on a barbeque spit.

Humans of ancient times did not understand the movements of the earth. They thought the sun orbited the earth, rising in the east and setting in the west. It was not until the mid-nineteenth century that proof of the earth's rotation was found. A Frenchman named Foucault affixed a heavy iron ball to a long steel wire that was, in turn, fastened to a ball joint in the ceiling of a high-domed cathedral in Paris. Then, he pushed the ball and started it swinging to and fro. He had theorized that a pendulum would not change its course unless it were acted upon by some external force. He set up little wooden pegs on the floor around the pendulum. The swinging iron ball began to knock them down, each in turn. At first he thought the pendulum had altered its direction—then he realized that the pendulum had not swerved from its path, but the earth had rotated beneath the pendulum. The iron ball appeared to rotate clockwise in its arc, which actually meant that the earth

Foucault's Pendulum.

63

was doing just the opposite: The earth was rotating from west to east, a counterclockwise direction for the Northern Hemisphere.* Further experiments showed that the nearer one was to the poles, the better the swinging pendulum worked at proving the earth's rotation. The reason for this is simple: At the poles, the pendulum swings back and forth over a rotating surface, while on the equator the surface of the earth rushes underneath in a straight line.

Today, we can observe the path of weather and communication satellites as they orbit the earth. Their paths appear to wander over the earth's face (Fig. 3-4), but actually, this meandering movement is illusory. The satellites follow a constant great-circle path around the earth (that is, their direction is steady and undeviating),

*For further information regarding Foucault and his pendulum, consult W. E. Johnson (1907), *Mathematical Geography* (New York: American Book Company), pp. 54–57. Also, if your college does not have such a pendulum, visit your nearest observatory, such as the one in Griffith Park, Los Angeles, or the one in Golden Gate Park, San Francisco.

FIG. 3-4 Path of a Satellite.   This diagram represents the movement of the earth to the east and the apparent "westward" movement of a satellite each time it passes around the earth.

Northern Hemisphere   Southern Hemisphere

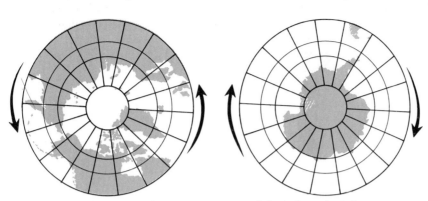

FIG. 3-5 **The Earth's Rotation about Its Axis as Viewed from above the Poles.**

but the earth is rotating, moving, beneath them, so the satellite does not pass over the same places each time it circles the earth. For example, a satellite crosses over the city of Honolulu, Hawaii, at noon, but 45 minutes later, as it again sweeps across the Pacific Ocean, it passes west of Honolulu. Why? Because the earth—and Honolulu—have moved eastward. A point to remember: Looking down over the North Pole would enable you to see that the earth rotates in a counterclockwise direction (Fig. 3-5). But, if you were looking down over the South Pole, the rotational direction would be clockwise. Always, however, the earth rotates from west to east.

EARTH MOVEMENTS AND THE SEASONS   Let us once again consider the earth's relationship to the sun. We know that the earth revolves around the sun and, at the same time, rotates around its own axis. These movements—revolution and rotation—coupled with an inclined axis, account for our changing seasons. As for the inclined axis, have you ever noticed that globes never stand straight up? Figure 3-6, which details the earth's orbit and the different seasons, shows that the earth's axis is tilted. Think, for example, of the outer edge of the earth's orbit as something like the edge of a large dinner plate, with the sun right in the middle of the plate. That plate, that path of the earth's orbit, is termed the **plane of the ecliptic.** The earth's axis is inclined, or tilted, away from that plane by an arc of 66.5 degrees (66°30′). Some prefer to think of the tilt as the inclination away from a line perpendicular (at right angles) to the plane of the ecliptic, to the amount of 23.5 degrees (23°30′). This inclination causes different regions of the earth to receive different amounts of solar energy, depending on the time of the year and the latitudes of the various regions.

From around March 21 until September 21, the Northern Hemisphere is tilted toward the sun, and consequently that hemisphere receives more of the sun's direct radiant energy than does the Southern Hemisphere. During the Northern Hemisphere's summer solstice (June 21), the region above the Arctic Circle has daylight for 24 hours, and the days are longer all over the Northern Hemisphere. Exactly

FIG. 3-6 The Earth's Orbit and the Plane of the Ecliptic.

the opposite is true for the Southern Hemisphere; nights are longer, and the area south of the Antarctic Circle has 24 hours of darkness. On December 21, the seasons are reversed. The Southern Hemisphere is then tilted toward the sun, and it is summer south of the equator and wintertime in the north.

During the periods of the **solstices,** the vertical, direct rays of the sun are over either the Tropic of Cancer (latitude 23°30′ north) on June 21 or the Tropic of Capricorn (latitude 23°30′ south) on December 22. Remember: The two hemispheres are exactly opposite during the solstices, the Northern Hemisphere's summer is the Southern Hemisphere's winter, and vice versa. Also, the days are longer and the nights shorter during the time of the summer solstice, and the opposite during the winter solstice. Only the equator has days and nights of equal length throughout the year.*

Between the times of the two solstices, at the midway points in the earth's journey around the sun (about March 21 and September 21), neither hemisphere receives more direct solar energy than the other. The earth is still tilted, but the inclination is at right angles to the plane of the ecliptic, so both hemispheres have 12-hour days and 12-hour nights. This is the time of the **equinoxes:** March 21 is called the *vernal* (or spring) equinox in the Northern Hemisphere and the *autumnal* (or fall) equinox in the Southern Hemisphere. On September 22, it becomes the autumnal equinox for the north, the vernal for the south.

Between the solstices and the equinoxes, the length of day increases or decreases, depending on the hemisphere, and the nights decrease or increase in length in corresponding opposition. If you were an astronaut out in space, looking down at the earth, you would see that one-half of the earth is always in shadow, while the other half is always in daylight. Of course, the rotation of the earth prevents any one side of the earth from getting too much solar radiation. The line between day and night, the line that divides the earth into dark and light is called the **circle of illumination,** or the *terminator* (Fig. 3–7). The line passes through the poles during the equinoxes, paralleling the meridians as the earth rotates. During the solstices, the line is inclined 23°30′ from the polar axis, making an angle of 66°30′ with the parallels of latitude.

How do the changing seasons affect humans? Those who live near the equatorial regions scarcely know what changing seasons are, but think of the people who live in Bodø, Norway, or Pt. Barrow, Alaska—lands where summertime means daylight for almost 24 continuous hours and wintertime means night for nearly 24 hours. In June, they get up in the morning with bright daylight all around, and when they go to bed at night, there is still daylight. If a person is psychologically attuned to going to bed when it is dark, what do they do when there is no night? And consider what happens to their nervous systems six months later, when there is little or no daylight? Some people get depressed during these long winters, and suicides are common during this period.

---

*The term *solstice* comes to us from the Latin *solstitium: sol* (sun), *stitium* (stands still); therefore: a time when the sun seems to remain motionless in the heavens. *Equinox* comes from the Latin *aeques* (equal) and *nox* (night); thus: a time when days and nights are of equal length.

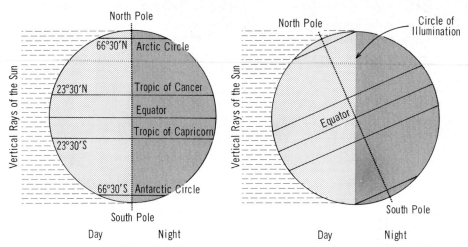

EQUINOXES   SOLSTICES

FIG. 3-7   The Circle of Illumination.

From the high northern latitudes, the sun appears far away to the south and relatively low to the horizon. During the summer, twilight seems to linger because of the low horizon angle of the sun, but this is not the case for areas close to the Tropic of Cancer, such as the Hawaiian islands. In Honolulu, the sun appears to be almost overhead throughout the year, and when sunset comes, it comes with great abruptness: If you wish to photograph the brilliant colors of a Hawaiian sunset, you should be ready and waiting, for it all happens, seemingly, in a matter of seconds.

Earth's Power Plant: Solar Energy

The sun is more than just a gigantic ball of flaming gases hanging out in the center of our solar system. It can be likened to an enormous furnace that is constantly emitting massive amounts of energy in every possible direction (Fig. 3–8) in the form of visible light rays (41 percent), infrared and heat rays (50 percent), and ultraviolet, gamma, and X rays (9 percent). Fortunately for us, however, our earth is at the correct distance from the sun for its size to receive the energy we need. In addition, we have our atmosphere to protect us from excessive doses of solar energy.

SOLAR ENERGY AND THE EARTH'S ATMOSPHERE   Consider what might happen if our atmosphere did not exist and incoming solar radiation could strike the earth's surface without interference. Although it is unlikely this will ever occur, we had advance warning of what could happen when humans learned how to manufacture their own form of radiant energy—the atom bomb. When a nuclear device is

Summer Solstice, Bodo, Norway. This series of photos taken at Bodø, Norway (67° north latitude), demonstrates that the land of the midnight sun is not a figment of the nordic imagination.

exploded, the nearby earth is showered with deadly gamma rays, X rays, ultraviolet rays, as well as visible light and heat rays. If people are too close to the scene of the blast, they might be blinded by the intense light, burned by the heat, scarred by the gamma rays. If they are not in the near vicinity, but happen to be downwind, they might not be instantly killed but will experience scalding and burns (as did the Japanese fishermen who were caught unawares too close to Bikini Atoll when that great test bomb was exploded).

Surrounding the earth and extending far out into space to a distance of approximately 50,000 miles (80,000 kilometers) is the earth's magnetic field. Within this vast region (often referred to as the *magnetosphere*), there is another form of solar energy—**corpuscular energy.** This energy issues from the sun as clouds of electrified

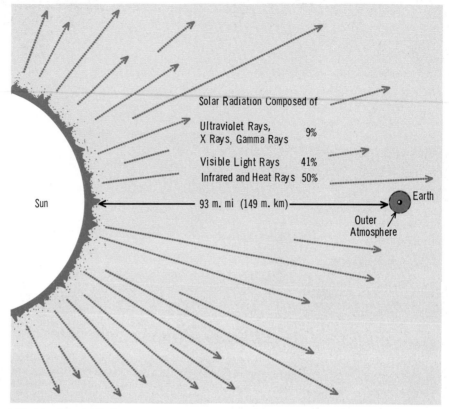

Solar Radiation Composed of

Ultraviolet Rays,
X Rays, Gamma Rays   9%

Visible Light Rays   41%
Infrared and Heat Rays 50%

Sun

93 m. mi (149 m. km)

Earth

Outer
Atmosphere

FIG. 3–8   Solar Radiation.

particles that appear to have a direct effect on television and radio communication, as well as on the earth's magnetic field. Also known as "magnetic storms," they occur about once a month, gushing forth from the sun to spray our entire solar system with tiny molecules of helium, hydrogen, oxygen, and nitrogen. Our earth, being a magnet, catches a certain percentage of these solar particles at each pole. The particles cause friction with the nitrogen and oxygen in our atmosphere, creating vivid displays of green and red luminescence, which skip across the polar skies, giving us the curtains of light we call the *aurora borealis* at the North Pole and *aurora australis* at the South Pole.

It is possible that these solar particles could cause severe problems for us, but there are two belts composed of high-energy radiation particles (protons and electrons) that encircle the earth and protect us by absorbing much of the incoming corpuscular energy. The belts—called the *Van Allen radiation belts,* after the scientist who discovered them—are positioned at two elevations around the earth: 2500 miles (4000 kilometers) and 10,000 miles (16,000 kilometers).* The radiation contained

*To learn more of the discovery of these radiation belts, see James A. Van Allen (March 1959), "Radiation Belts around the Earth," *Scientific American* (offprint no. 248).

within these belts (Fig. 3–9) could be dangerous to astronauts attempting to go into space—interfering with their electronic gear, radar equipment, radios, telemetric equipment, and compasses. However, knowledge of the possible dangers beforehand enabled NASA to move ahead with confidence, and our astronauts have been able to navigate out through the "holes" in the doughnut-shaped rings above the poles as they journeyed to the moon.

Thus far we have examined the part played by the sun in controlling earth's movements, and we have seen how the earth's atmosphere is affected by the incoming solar energy. Let us now investigate what happens to the incoming solar radiation (usually abbreviated to one word, **insolation**) after it reaches the earth's surface.

SOLAR ENERGY AND THE HEAT BALANCE   As we have seen, the earth and its atmosphere receive only a fraction of the total emission by the sun. We have learned also that much of the original solar radiation, particularly the ultraviolet, gamma, and X rays, has either been deflected back into outer space or has been trapped at the outer levels of the atmosphere. Only those visible light and infrared rays that

Aurora Borealis.

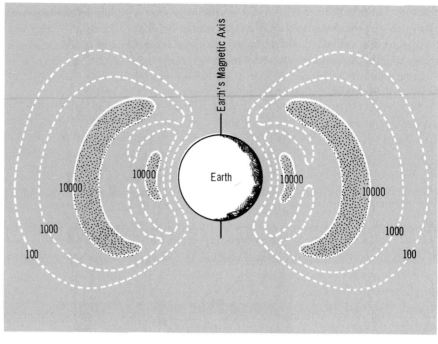

Earth's Magnetic Axis

Earth

10000   10000   10000   10000

1000   1000

100   100

  Force fields

  Radiation belts

FIG. 3-9   Van Allen Radiation Belts. High-energy protons and lower-energy electrons that create a penetrating type of radiation make up Van Allen radiation belts (and their accompanying force fields). The intensity of radioactivity in the various belts and force fields is indicated by numbers (measured by a Geiger radiation counter in "counts per second"). (Adapted from U.S. Navy, "The Upper Atmosphere," U.S. Navy Weather Research Facility Chart NWRF 26-1161-051, Washington, D.C.: U.S. Government Printing Office.)

have escaped such deflection or entrapment are left to continue the journey toward the earth. By the time the surface of the earth is reached (Fig. 3–10), about 35 percent of the original incoming solar radiation is reflected into space, 19 percent has been or will be absorbed by the lower atmosphere, and the remaining 46 percent is or will be absorbed by the earth itself.

Does this mean that the 65 percent of solar radiation that comes into our atmosphere or reaches the earth is retained permanently? Obviously, the answer is no, for if this were so, the temperature of the lower atmosphere would rise to a level that might make life on earth impossible. Although retention of solar radiation has occurred many times throughout the history of the earth, in relatively recent times a state of equilibrium has been established—a kind of balance in which the earth sends back into outer space an amount of radiant energy equal to the amount received. This is called the **heat balance,** and many factors go into its maintenance. The earth and its atmosphere must set into motion various processes of heating and cooling, such as radiation and reradiation, compression and expansion, condensation and evaporation, conduction, and methods of energy transfer, such as convection and advection. In addition, there are a number of controls that exert influence on the heating and cooling processes.

CONTROLS OVER HEATING AND COOLING Insolation is not received equally by the various levels of the atmosphere or by different places on the earth. Figure 3–11 illustrates some of the obstacles insolation must overcome before it can get on with the task of supplying the earth and the atmosphere with heat. By the time the tropopause has been reached, much of the insolation has been either absorbed or transformed into other forms of energy in its passage through the thermosphere,

FIG. 3–10 Incoming Solar Radiation (Insolation).

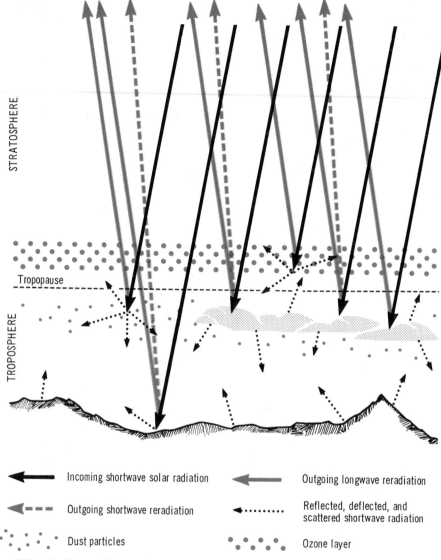

STRATOSPHERE

Tropopause

TROPOSPHERE

◄━━━	Incoming shortwave solar radiation	◄━━━	Outgoing longwave reradiation
◄━ ━ ━	Outgoing shortwave reradiation	◄┄┄┄	Reflected, deflected, and scattered shortwave radiation
Dust particles		Ozone layer	

FIG. 3–11 **Energy Movement.**

mesosphere, and stratosphere. Here, the controls are the deflection and the scattering of solar particles as they encounter other particles of matter. Deflection is important in the troposphere, too, but a different term is used—**albedo.**

Albedo Basically, albedo means reflection of radiant energy by an object, expressed as a percentage. The percentage for heat or visible light rays is much higher for a snowfield than it is for a dark pine forest. A dark item such as a black suit would have a lower albedo than a white suit, which you can easily prove on a warm summer day. Insolation would be readily absorbed by the dark clothing, thus heating you considerably, but the white, or light-colored clothing, would reflect the insolation and you would feel cool. Meteorologists have calculated the albedo for different substances and find that clouds and regions of ice and snow consistently have the highest albedo (up to about 80 percent), grasslands are intermediate (about 25 percent), and oceans and other water bodies have the lowest (about 5 percent). The earth, as a whole, averages an albedo of around 40 percent. It might be noted that if the energy arrives at a low glancing angle, the albedo increases. One can easily see why, in places near the equator, the albedo would be higher during mornings and afternoons than at noon, and why the albedo would be higher on polar snowfields than on equatorial Mt. Kenya.

Land and water differences   The differences in the absorptive qualities of land and water constitute important controls in the heating and cooling of the earth, but land and water do not absorb insolation equally. Oceans heat up very slowly, much more slowly than does the land, but the heating goes much deeper. Whereas insolation will be absorbed only a few inches down into soil, it will penetrate as much as 6 feet (2 meters) in tropical waters. Consider the difference in the warmth of sand alongside an inland lake or near the ocean and the water itself. If it were in mid-summer, at noon or early afternoon (when the insolation was at its highest), the sand would be hot to touch, but if you wiggled your toes an inch or two into the sand, you would discover that the sand was cold there. If the temperature of the sand were, say, 90°F (32°C) and the temperature of the body of water were near 70°F (21°C), you would have no difficulty noticing the difference—the water would seem very cold to you. But what happens when the sun has set? The sand loses its heat rapidly and feels cold. The lake or ocean does not lose heat as rapidly, and its temperature falls perhaps 2 or 3 degrees. Compared to the sand, the lake or ocean would seem warm.

There may be differences in surface temperatures in winter and summer, but generally, the oceans and other large bodies of water are so stable that they are a moderating influence on nearby land. Contributing factors to this, in addition to the reason discussed above, include the movement of warm and cool ocean currents, the upwelling (convection currents) of cool water near hot lands, evaporation that continues all year (and is especially appreciated during the hot summer months), the great size and depth of the oceans, and the immense volume of water near the equator that warms and retains its heat for long periods.

As an example, compare the differences in monthly temperatures between two cities that are positioned at approximately the same latitude (Fig. 3–12), but where one is located next to the ocean and the other is inland. What if you were overly sensitive to cold weather and someone told you that Nashville would be a better place for you to live because its average temperature for the year is higher than San Francisco? Would you be getting the entire story? Nashville's average temperature is 59.5°F (15°C) and San Francisco's is about 56.5°F (13.6°C). If you consult the monthly average temperatures, you will note that Nashville's winter temperatures range from 40°F (4°C) to about 45°F (7°C), while winter in San Francisco averages

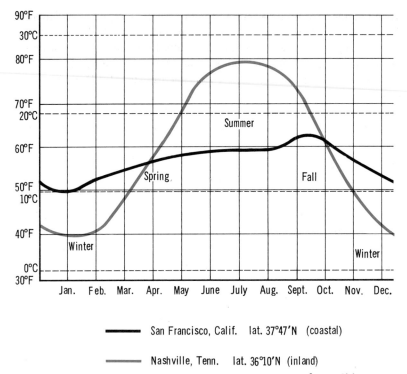

FIG. 3–12 Influence of Land and Water on Air Temperatures of Two Cities at about the Same Latitude.

about 55°F (12.8°C). Consider also the annual range (from the coldest to the warmest months), and you will find that Nashville has a variation of 40°F (4°C) and San Francisco has a range of only about 12°F (−11°C). Thus, you might decide that San Francisco's temperature stays fairly even all through the year, while Nashville is subject to extremes.

Latitude Another important control over an area's receipt of insolation is latitude. When we discussed the yearly orbit of the earth around the sun and the resulting changing of the seasons, it was pointed out that the earth's axis is tilted to the plane of the orbit. Thus it is obvious that not all areas on the earth's surface receive the same amount of insolation. Even during the equinoxes, when neither the Northern or Southern Hemisphere is tilted toward or away from the sun, insolation is not the same for all regions.

Consider Fig. 3–13 and contrast the possible receipt of insolation for points X, Y, and Z during the solstices and equinoxes. We can see that the Northern Hemisphere is in its summer solstice (*a*) and point X would have a longer day; thus its receipt of insolation would be considerably greater than at point Z.

The opposite occurs during the winter solstice (*c*). During the equinoxes (*b*), both X and Z receive the same amount of insolation. But how does point Y compare? As it is only 10 degrees south of the equator, its receipt of insolation will be almost constant throughout the year. An important factor is the angle of impact of insolation. For point Y, the angle is nearly vertical, while for points X and Z, the solar rays will strike at an oblique angle, with resulting high albedo. In addition, insolation coming in at an angle must travel through a greater amount of atmosphere, and when it strikes the earth's surface it must cover a greater area (Fig. 3–14), thus losing much of its intensity.

Albedo, the time of day, the season of the year, as well as latitude, play important roles in determining the amount of insolation any area will receive, but there are other controls. These are, for example, the presence in some areas of great mountain ranges, broad deserts, heavy forests, icecaps and snowfields, as well as powerful wind systems.

HEATING PROCESSES   Earlier in this chapter, we learned that incoming solar radiation warms the atmosphere very little as it passes through it. The main heating processes are performed on or near the earth's surface, and they include radiation (and reradiation), conduction, compression, and condensation. It will become evident,

FIG. 3–13 **Influence of Latitude and the Seasons.**   (*a, c*) solstices; (*b*) equinoxes.

(a)

(b)

(c)

FIG. 3–14 Rays of the Sun and Latitude. Assume a specific quantity of insolation. The rays of the sun strike the higher latitudes of the earth at oblique angles (*a*). They must therefore cover a greater amount of earth surface and atmosphere than do rays that strike the surface at or near the equator (*b*). Vertical rays are concentrated on a smaller amount of earth space and travel through less atmosphere.

as we proceed, that some of these heating processes also perform the task of cooling the earth and modifying the extremes of temperature that otherwise might result.

Radiation and reradiation The primary heating processes are **radiation** and **reradiation,** the result of the absorption of incoming shortwave solar energy by the earth and the subsequent reradiation of longwave energy into the atmosphere.* Because the production of shortwaves requires a very hot radiating body, the earth can only produce forms of longer length energy waves. This longwave radiation is sometimes referred to as *infrared, ground,* or *terrestrial* radiation. When the ground is warmed under the impact of insolation, the molecules become excited and begin to vibrate rapidly, causing the release of the infrared radiation. These waves rise rapidly into the lower atmosphere, where about 90 percent of them are trapped and absorbed by particles of carbon dioxide, ozone, and water vapor. Eventually, the atmosphere radiates longwave radiant energy back toward the earth, warming the atmosphere by its passage.

This process by which our earth and atmosphere are heated and warmed has been likened to what happens when you leave your car out in the sunlight with the windows closed. It gets hot inside. Why? Because shortwave solar energy can easily penetrate the car's windows, and once inside the car it is absorbed by the upholstery and reradiated as long waves—which cannot escape the car's interior. Obviously, your car heats up.

*Both shortwave and longwave radiant energy travel at the speed of light—186,000 miles (298,000 kilometers) per second. However, shortwaves are narrower and can squeeze between gaseous particles in the atmosphere, while longwaves take up more space and are constantly being intercepted by dust, gaseous particles, and especially, water vapor in the lower atmosphere.

Another analogy would be a greenhouse—a glass-enclosed building where, even in cold climates, one can grow tropical plants. Here again, the powerful shortwave energy passes through the panes of glass and is absorbed by the plants, clay pots, soils, wooden benches, and the earth and is reradiated into the enclosed space. The longwave radiation cannot escape through the glass and is deflected back and forth within the greenhouse. As this process (Fig. 3–15) continues, the temperature inside the greenhouse builds up to, perhaps, a nice 70°F (21°C)—even though the outside temperature may be a very cold 40°F (4°C). Meteorologists have long referred to this process as the **greenhouse effect,** but it must not be inferred that the panes of glass actually absorb the longwave radiation and then reradiate it back down. It is truly a deflection process, and not exactly like that which occurs within our

FIG. 3-15 The Greenhouse Effect. The length of an energy wave depends on the heat of its source. Solar energy waves are strong, short, and move with great intensity and speed. Heat that is reradiated from the earth or objects such as pottery, vegetation, and people is warm, not hot, and the resulting heat waves are weaker, longer, and slower. Infrared and X-ray photography, in which the heat emanating from a body (or the bones within) is photographed, are partly based on this principle.

 Incoming shortwave solar radiation penetrating glass panes of greenhouse

 Reradiated long waves issuing from plants, soil, and wood inside the greenhouse

atmosphere. It is becoming more common today for meteorologists to refer to the warming of our earth and atmosphere by radiation as the *atmosphere effect,* rather than as the greenhouse effect.*

What would happen if the atmosphere became too dense and did not allow 10 percent of the longwave radiation to escape to outer space? Temperatures within the atmosphere and on the earth would increase to a point of high humidity and extremely high temperatures. Earth would have a tropical age. Water from the oceans, lakes, and streams would evaporate excessively and form clouds at the tropopause. This newly formed cloud layer would prevent incoming solar radiation from penetrating our atmosphere, and the air temperature would drop, bringing on an ice age. Earth scientists have assured us that there have been four ice ages, alternating with interglacial periods, during the last 1 million years. Can we predict when this cycle might occur again? The last ice age ended some 10,000 or more years ago.

Conduction Another process of heating is **conduction**, in which, by direct contact, heat will move from a hot or warm object to a colder one. Some substances are good conductors, others are not. Wood is a bad conductor, but an iron bar or a soup spoon is a good one. Air is a poor conductor, and as a result, cold air settling over warm earth will heat up only a few centimeters from the surface. A lake or ocean warmed by a summer sun will heat up to great depths and will consequently be able to warm the bottom layer of a settling cooler air mass for a relatively long period of time by the simple passage of warmth from the water to the air. Land, on the other hand, does not heat up to any great depth. Further, land loses most of its heat by radiation, and what is left to warm cooler air via conduction is minimal.

Compression The heating process of **compression** occurs when an air mass descends from a higher to a lower elevation and gains in temperature as it moves downward (Fig. 3–16). The warming rate averages about 5.5°F for every 1000 feet of descent (1°C per 100 meters). Inhabitants of Samarkand (Soviet Uzbekistan) are constantly being reminded of this process by the air masses blowing down from the Pamir Mountains of Afghanistan. These hot, dry winds, called the **Afghanets,** carrying great clouds of fine dust particles, blot out the southeastern horizon and make life for the Uzbeks rather uncomfortable.

Residents of Santa Ana, California, also feel the warming effects of descending air masses, especially during winter months, when these **Santa Ana winds** move down from the Great Basin region of Nevada and Utah.† These winds are the result of the settling of air that moves poleward from the equator at high altitudes. As

---

*For an interesting explanation of the actual workings of a greenhouse and our atmosphere, see Robert C. Gleage and Joost A. Businger (1963), *An Introduction to Atmospheric Physics* (New York: Academic Press), pp. 153–154.

†*Santa Ana* (pronounced *Sant-ana*) is the correct term for this local winter heating process. It was given its name in the seventeenth century by Spanish explorers who were puzzled by 70°F (21°C) temperatures in midwinter in the region south of present-day Los Angeles. Similar warming winds are to be found in many places: *chinook* for the region east of the Rockies; *foehn* in Switzerland and Europe; *samoon* in Iran.

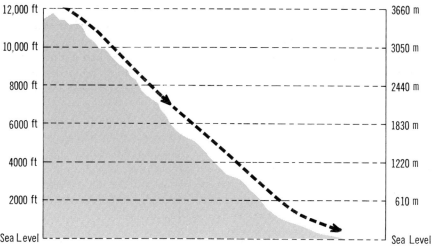

FIG. 3–16 Heating of Air by Compression. The temperature of an air mass changes with differences in altitude. If the air is still, nonmoving, the normal lapse rate (or change) is 3.5°F per 1000 feet (0.65°C per 100 meters). If the air mass is moving down the side of a mountain, or from a higher elevation to a lower one, the change increases to about 5.5°F per 1000 feet (1°C per 100 meters) as it descends. In the illustration, an air mass descending from 12,000 feet (3660 meters) increases its temperature 66°F (36.6°C) by the time it reaches sea level. On nights when the skies are clear and the air pressure is low, cold air will form close to the ground, and if there is a slope, the heavier, colder, denser air will flow beneath any warmer, lighter air present, producing a cooling wind that is called *katabatic* (or mountain wind). This is a form of air drainage.

the warm air cools, becoming heavier and denser, it begins to sink, spiraling outward from its central area in a clockwise motion. The moving air cannot remain in this plateau region. It must move out, gravitating down from the plateau toward the Pacific Ocean, blowing southwesterly down through mountain passes and across the Mojave Desert, descending from an average elevation of about 6000 feet (1800 meters) to the near sea-level elevation of the Los Angeles region. The descending air mass gains about 33°F (18°C) in its downward passage, and if, for example, it began in Nevada at a cool 55°F (12.8°C), its temperature would be increased to about 88°F (31°C) on arrival in Los Angeles. In addition, the funneling process (being squeezed between mountains) causes these drying winds to be quite gusty. Brush and forest fires are particularly frequent in Santa Ana or Afghanets areas. This warming of moving air is called **adiabatic heating,** which means that the air mass neither gains nor loses any of its energy.

Condensation In the heating process of **condensation,** water in a gaseous state (as vapor) is converted to a liquid (water) or a solid (ice or hail) state and heat (called *latent heat*) is released into the air mass. This occurs when a warm air mass is lifted to a cooler and higher elevation or moves over a cold landmass at a low elevation, then cools until it can no longer retain the water it holds in its gaseous state. This level, called the **dew point,** is when condensation results, when moisture-laden air reaches complete saturation. The point at which this occurs varies with the temperature of the air mass: The warmer the air, the more water vapor it can retain; the cooler the temperature, the less it can hold. Any excess must be released. It must be noted, however, that not all condensation results in rain, snow, or hail. When dew point is reached, moisture might become fog, dew, or frost (if the air temperature is at or below freezing).

COOLING PROCESSES   The temperature of the atmosphere would rise to uncomfortable levels in most regions were it not for a number of cooling processes. Some of the previously discussed heating actions also produce cooling. For example, when heat is radiated from the land or waters, the temperature of the radiating object or substance is reduced in equal measure. The same applies to the process of conduction when, by direct contact, heat is transferred from one object to another. For example, when cold air overlies a warm body of water, heat will move from the water to the air, and the water will lose temperature. If fog or rainfall result from condensation, the land on which this form of moisture falls will be cooled, as heat is transferred to the moist air mass. Some of the more significant cooling processes are the result of evaporation, expansion, and advection.

Evaporation   In the process of **evaporation,** the opposite of condensation, water in a liquid or solid state is transformed into a gas (water vapor). Rather than releasing heat (like the latent heat of condensation), the surrounding air mass loses heat as the molecules race wildly about, colliding with each other, finally breaking free to take some of the water's energy with them. If you have ever exercised on a hot, dry day, you know that the perspiration emitted by your body is immediately taken up by the air, and you feel cool. If the air were hot but sticky (meteorologists say **humid,** meaning that the air around you is almost to the saturation point, and thus cannot accept more moisture), the beads of perspiration would stay on your body, and you would not feel cooled. If, however, a breeze were blowing, the ability of the air to accept more moisture would be increased, and you would feel the cooling effect of the wind. We might say that evaporation depends on a number of factors: the presence or absence of wind, the aridity (dryness) of the air, and the temperature. On a windless, midsummer day, people in humid Jackson, Mississippi, might be more uncomfortable than people in hot and dry Phoenix, Arizona.

Expansion   The process of **expansion** is the opposite of compression. As an air mass will warm when the molecules are pressed more tightly together, so the air mass will cool if the molecules are no longer confined in a small amount of space and can move about more freely. This has nothing to do with outside temperatures. There is no transfer of heat by such processes as conduction or radiation to the surrounding air. It is purely automatic and is made possible because of differences in air pressure over an area.

This is how expansion occurs. Insolation warms the earth, causing the overlying air to be heated. As the air mass warms, it expands and is lighter in density. It begins to rise, and as it rises, the pressures surrounding it become weaker, thus allowing the molecules within the rising, warm-air mass to separate and move further and further apart. Meteorologists call this form of cooling **adiabatic cooling.** The upward movement of the expanding air mass is caused by an energy transfer process called **convection** (vertical movement or uplift of an air mass), and the rate (amount) of cooling is referred to as the **lapse rate.**

There are three types of lapse rates—normal, dry adiabatic, and wet adiabatic—but only one has anything to do with quiet, unmoving air. This is the *normal* lapse rate. In this case, the change in temperature is strictly regulated by differences in altitude. As was discussed earlier, the atmosphere is densest near the earth's surface. Obviously this means that it is less dense at higher elevations. The rate of decline is about 3.5°F for every 1000 feet of altitude increase (0.65°C per 100 meters). Suppose you wanted to test this by taking a thermometer with you on a balloon ride up to 10,000 feet (3000 meters) above the earth. If the temperature of the air at sea level were, say, 90°F (32.2°C), it would be about 17.5°F (9.75°C) cooler at the 5000-foot (1500-meter) mark, and 35°F (19.5°C) cooler at the top of your ascent, where the air temperature would be 55°F (12.7°C).

When an air mass is moving, the lapse rate increases, because the act of movement causes the air mass to lose additional heat. The heat loss now becomes 5.5°F per 1000 feet of altitude increase (1°C per 100 meters). This is termed the *dry adiabatic* lapse rate. If the warm air mass is also heavily laden with moisture and the convectional uplift moves it to a height where the loss of temperature decreases its moisture-carrying capacity, condensation may occur. If this happens (remember our discussion on page 81), latent heat is produced, and the heat loss is diminished to a lapse rate of only about 3.2°F per 1000 feet (0.6°C per 100 meters). This is termed the *wet adiabatic* lapse rate. For an example of these lapse rates in action, see Fig. 3–17.

Advection   The horizontal movement of air, **advection,** also is an important cooling process. Cool or cold air is heavier and denser than warm air, and as heated air rises from its heating surface, the heavy air slides into what has become an area of low air pressure. Residents of the American Midwest can testify to the welcome intrusions of cool air from Canada that occasionally temper their long, hot summers. This can also happen along the shoreline of a lake or ocean, when cool winds blow inland to replace warm air rising over towns and cities. It must be noted, however, that advection can work the other way, too, as when a warm air mass glides north from the Caribbean or the Gulf of Mexico to plague an already overheated southeastern United States with additional humidity and, sometimes, hurricanes.

Temperature inversions   The usual situation is for temperatures to be cooler the further away from the earth's surface one gets. However, what happens in the heating and cooling processes and to the lapse rates of air temperatures when the reverse situation occurs—temperatures increase, rather than decrease, with altitude? Meteorologists call this a **temperature inversion.**

Static and dynamic inversions   There are two kinds of temperature inversions: static and dynamic. **Static** inversions usually are brought on during long winter nights

83

FIG. 3-17 Adiabatic Lapse Rates. In this diagram, a warm air mass moves against a mountain and is forced to rise to higher elevations. If its temperature at point A were 80°F (27°C) at sea level, and if the dry adiabatic lapse rate were in effect (and not interfered with by snow or rain), the temperature of the rising air mass would decline to 36°F (2.2°C) at point B, the 8000-foot (2440-meter) elevation. If condensation were to occur at point B and continue to the top of the mountain, point C, the temperature of the air mass would decline to 29.6°F (−1.3°C).

that are cool, clear, calm, and dry. They occur generally during high-pressure conditions that permit rapid radiation of heat from the earth. If there is little air movement (either vertically or horizontally), rapid ground radiation is given off, and the earth becomes cooler than normal. The air in contact with the ground becomes chilled by conduction as it touches the cold ground and, therefore, loses heat. The resulting cold layer of air acts as an effective barrier against any intrusion from above by warm air. *Dew* will result with a lowering of the temperature. If the temperature drops below freezing, *frost* will occur. If the air mass is damp, a *radiation fog* will occur in some valley areas, such as in California's San Joaquin Valley.

Dynamic inversions are created when a mass of heavy and cold air settles over an area, sliding down the sides of mountains and hills, forming great pools of cold air on the valley bottom, effectively shutting off the people and the crops on the ground from any warm air aloft. Farmers often try to take advantage of the movement of the cool air down the slope (a process called *air drainage*) by planting their crops on the hillsides where the moving air does not linger long enough to chill the crops. Farmers who do use the bottom lands often try to protect their crops by using smudge pots to raise the air temperatures, by implementing nighttime irrigation, or by installing wind fans. In seriously plagued areas, farmers often plant cotton, sugar beets, or feed crops such as alfalfa—crops that can thrive in cold air.

Temperature inversions and the urban environment What, if any, are the adverse effects of inversions? If an inversion does exist, what happens when a city erupts into its frenzied daytime activity, when people awake to go about their daily tasks? At first, the sky is clear and the air is cool and easy to breathe, then the automobiles, trucks, buses, trains, jets, and factories begin their production of various gaseous mixtures (such as carbon monoxide, sulphur dioxide, and others), which unite with ozone particles in the atmosphere. The resulting pollution begins to build up, and the noxious **smog** is born.

As the city heats up from its receipt of insolation, the gaseous mixtures rise, but they cannot break through the upper levels of the inversion layer, so they thicken and spread back to the earth's surface. The inhabitants of the city—whether it is Los Angeles, New York, or Russia's Moscow or Irkutsk—begin to have a number of problems. They may experience stuffy noses, watering eyes, aching lungs, and they find that distant mountains can no longer be seen. Citizens of Los Angeles often joke about the "good old days" when "on a clear day you could see Catalina Island."

Why are smog-producing temperature inversions relatively common in urban areas?* One meteorologist explained that cities develop inversions because they tend to be warmer than the surrounding countryside. There are a number of reasons for

85

*For a thorough accounting of the effect of cities on the environment, see: William P. Lowry (August 1967), "The Climate of Cities," *Scientific American* (offprint no. 1215); R. A. Bryson and J. E. Ross (1972), "The Climate of the City," in T. R. Detwyler and M. G. Marcus (eds.), *Urbanization and Environment* (Belmont, Calif.: Duxbury Press), pp. 51–68; J. E. Peterson (October 1969), *The Climate of Cities: A Survey of Recent Literature,* Publication AP-59 (Washington, D.C.: U.S. Department of Health, Education, and Welfare, National Air Pollution Control Administration); J. R. Norwine (July 1973). "Heat Island Properties of an Enclosed Multi-Level Suburban Shopping Center," *Bulletin of the American Meteorological Society,* 54, no. 7, 637ff.

Smog Formation in Los Angeles.

this. First, there is a difference in surface materials: Concrete buildings, for example, heat to a greater depth and retain their heat longer than does soil. Second, buildings and other city structures absorb more heat because of their shapes and closeness to each other than do features of the natural landscape. Third, factories, automobiles, and residential heating systems generate a considerable amount of heat. Fourth, cities retain little fallen rain or snow, because most of such precipitation runs off through drainage systems, leaving little to evaporate and thus to function as a coolant.[1]

Basically, then, when warm air, for whatever reason, becomes separated from the ground by an intruding mass of cold air or, owing to radiation, by a rapid loss of heat at ground level, a temperature inversion is created.

Review and Discussion

1. Why is the earth's position relative to the sun considered important?

2. Explain these terms: *revolution, rotation, aphelion, perihelion.*

3. Explain the difference between tropical year and sidereal year.

4. Explain Newton's first law of motion: inertia.

5. Why do we have leap years?

6. Has the earth always rotated around its axis at the same speed?

7. What did Foucault prove with his pendulum?

8. Why do earth's weather and communication satellites appear to wander as they orbit the earth?

9. Which direction does the earth rotate: east to west or west to east?

10. What is the plane of the ecliptic?

11. Of what benefit to mankind is the inclination of the earth's axis?

12. What has the tilt of the earth's axis to do with the changing seasons?

13. Explain the difference between the solstices and the equinoxes.

14. What is the circle of illumination?

15. Why is the sun sometimes likened to a giant furnace?

16. How does our atmosphere protect us from excessive solar radiation?

17. What is corpuscular energy?

18. What are the Van Allen radiation belts?

19. What causes the auroras?

20. Explain the term *insolation.*

21. Discuss the significance of the earth's heat balance.

22. Explain the term *albedo*. Why is it higher at the poles than near the equator?

23. Explain why a large body of water (lake or ocean) can be an important modifier of a region's climate.

24. What does an area's latitude have to do with its heating and cooling?

25. Explain these terms: *radiation, reradiation, conduction, convection, compression.*

26. Explain the greenhouse effect.

27. What causes hot drying winds, such as Santa Anas, Afghanets, and chinooks, to occur?

28. Explain the difference between a normal lapse rate and an adiabatic lapse rate.

29. If an air mass were to descend from a mountain-top elevation of 15,000 feet (4500 meters), would its temperature increase or decrease by the time it reached sea level? What would be the amount of change?

30. If an air mass ascended the windward side of a 16,000-foot (4800-meter) mountain, with condensation occurring at the 12,000-foot (3600-meter) level, what would be the temperature increase or decrease at the top?

31. Explain these terms: *evaporation, advection, saturation, condensation, dew point.*

32. What causes temperature inversions?

33. Explain the difference between static and dynamic inversions.

34. How do temperature inversions affect life in urban environments?

35. What is smog?

Note

[1] William P. Lowry (August 1967), "The Climate of Cities," *Scientific American* (offprint no. 1215), pp. 15–23.

Additional Reading

Beiser, Arthur (1962). *The Earth.* New York: Time–Life Books.

Bergamini, David (1962). *The Universe.* New York: Time–Life Books.

Nicks, Oran W., ed. (1970). *This Island Earth.* NASA publication no. SP-2500. Washington, D.C.: Office of Technology Utilization, NASA. A valuable collection of photographs of the earth, taken by astronauts in space and by satellites.

4

EARTH'S DYNAMIC ATMOSPHERE

What is the atmosphere? Why, it is the envelope of air that surrounds our planet. It gives rise to the climates of the world; it gives birth to the weather that confounds us every day. Charles Dudley Warner once said: "Ah! the weather! Everyone complains about it, but no one does anything about it!" True enough for the first half of this century. But, true today? No.

Because earth has a dynamic atmosphere, it also has weather and climate. *Weather* and *climate* are two words that seem to mean the same thing and, to some people, are interchangeable. Yet, in a way, they might be likened to the terms *mother* and *daughter*—related, but not the same. In this sense, weather is the mother of climate.

Weather is "what it is like" today or tomorrow. It is the condition of the earth's atmosphere over a specific location for a specified period of time, such as a few hours or a couple of days.

Climate, on the other hand, is the weather that is characteristic of a place for a much longer period of time. The climate of a region is based upon a generalization of the averages gained from a systematic study of weather data (statistics) gathered over 30 or more years. Numerous elements comprise climate: temperature, moisture, atmospheric pressure, and wind. For example, while the climate records for Fresno, California, show that it averages 11.2 inches (28 centimeters) of rainfall per year, there has been no actual year when that exact amount has fallen.

A person would be concerned about an area's weather if he were planning a short vacation there, but he would be more concerned about the climate if he were planning to move there permanently. Weather deals with today or tomorrow—what the temperatures will be, what the chances are for rain, and if there will be wind, while climate goes deeper, entailing the evolution of the area's natural vegetation, soils, water supply, and landforms. Climate can tell us why one region on the earth's surface is a vast empty desert, another is a thick, heavy rainforest, and still another is a wide open grassland.

The study of weather is known as **meteorology** (from the Greek *meteora,* "things out there"), while the study of climate is called **climatology** (from Aristotle's *klimata,* "climate regions"). Both the meteorologist and the climatologist investigate and analyze the workings of the earth's atmosphere. They are concerned with the different elements and the way these are controlled or modified by differences in latitude; sizes, shapes, and distribution of landmasses and oceans; air temperatures; air pressures and winds; ocean currents; moisture and precipitation; altitude; and the presence or absence of mountain barriers.

In our investigation into why earth's atmosphere is so energetic, vital, and dynamic, we shall begin by looking closely at its structure, as well as at the different ways in which meteorologists and climatologists describe its composition.

Composition of Earth's Atmosphere

A number of ways have been devised to define the composition of our atmosphere. The most popular way was to set up two distinct regions: the "outer" and the "lower" atmosphere (Fig. 4–1). The outer atmosphere was composed of the **exosphere** (from

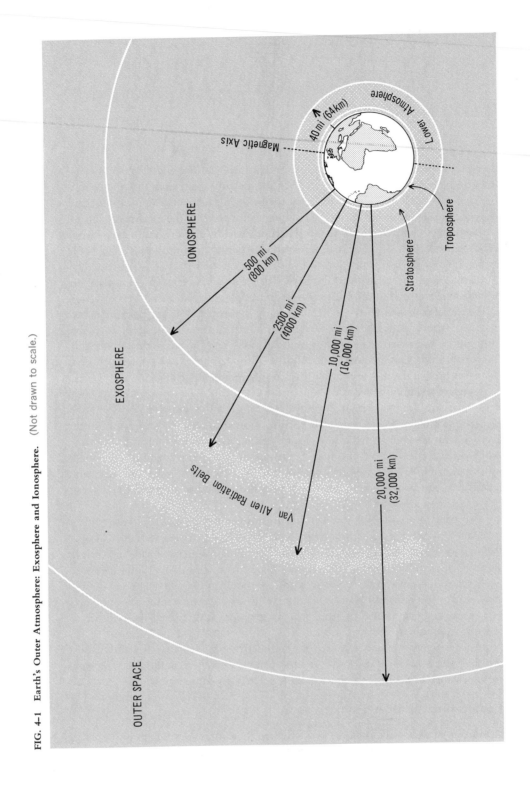

FIG. 4-1 Earth's Outer Atmosphere: Exosphere and Ionosphere. (Not drawn to scale.)

the Greek *exo,* "outer") and the **ionosphere.** The exosphere was supposed to be the outermost shell, or sphere, surrounding the earth, beginning at an altitude of about 650 miles (1040 kilometers) and extending outward to about 20,000 miles (32,000 kilometers). It was said to be a distinct part of the atmosphere, because it did contain charged particles of nitrogen and oxygen. It was difficult, however, to truly fit this region into the atmosphere, because no one really knew just where it came to an end. As a matter of fact, the exosphere just fades away into outer space.

The ionosphere is somewhat easier to explain, because it is an area of electrical activity—but again, it becomes confusing when one attempts to set absolute boundaries. Most **aeronomists** (scientists who study the upper levels of the earth's atmosphere) have been content to state that the ionosphere varies in depth and in distance from the earth. Generally, its lowest level begins about 50 miles (80 kilometers) above the earth and extends outward to about 650 miles (1040 kilometers). This region contains a great number of gaseous particles that are under constant bombardment by cosmic rays (solar particles). They become electrically charged, or *ionized,* and form into layers of ionized gases, which serve as a protective shield around the earth.

This shield serves mankind in a number of ways. For example, long-distance radio communication would be impossible if transmitted radio waves (which travel in a straight line) were not reflected (bent back) to earth by the ionized layers. Otherwise, they would escape into outer space. We can also be grateful to these ions for aesthetic reasons. As visible light rays bounce from ion to ion during their passage toward the earth, separation occurs, forming what we perceive as blue sky. It always seems bluest directly overhead, because when we look up, we only have to "see through" some 11 miles (18 kilometers) of atmosphere, whereas when we look at the sky over the horizon, we have to see through hundreds of miles of the densest and most polluted part of the atmosphere, which tends to lighten, or gray out, the blueness of the sky. From a health standpoint, also, we should be thankful for the ionosphere and its electrically charged particles, for they also intercept and deflect away from earth many of the incoming ultraviolet and gamma rays.

STRUCTURE OF THE ATMOSPHERE: GASEOUS COMPOSITION Another method of atmospheric classification uses the gaseous composition of different levels as the standard. Because the gases of the lower levels are well mixed and uniform in their proportions, the term **homosphere** (from the Greek *homo,* "the same") is applied (Fig. 4–2). Above the 72-mile (115-kilometer) level, however, the gases were found to be in distinct layers, and their proportions were quite different, so the term **heterosphere** (from *hetero,* "different") was used. Both regions are composed of the same kinds of gases (nitrogen and oxygen, plus minor amounts of other types of gases), but not in the same quantities or percentages. In the homosphere, the gas particles have sufficient density to form air, but in the heterosphere, they are widely separated and, for the tremendous space in which they circulate, insufficient in quantity to be dense enough to form air.

Air is mostly nitrogen (78 percent) and oxygen (21 percent) by volume, with the remaining 1 percent composed primarily of helium, argon, carbon dioxide, and

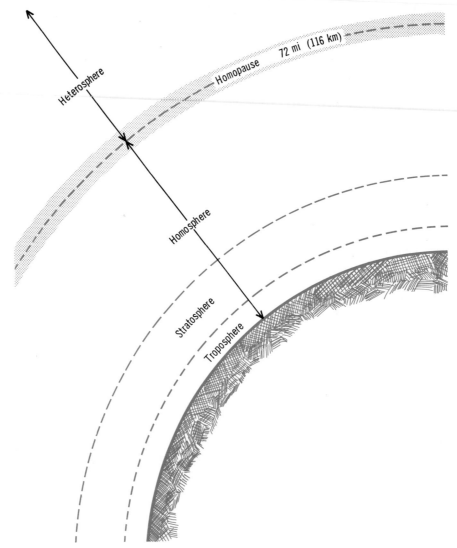

FIG. 4-2 Earth's Atmosphere, Classified by Gaseous Composition.

water vapor. It is colorless and odorless, and its very ingredients have helped to sustain life as we know it. Ninety percent of the mass of the atmosphere is contained in the lowest portion of the homosphere, a zone called the **troposphere.** This region might well be called the "home of mankind," for it supplies carbon dioxide for plants, disperses much of the pollutants we send into it, evens out the extremes of hot and cold, permits winds to carry moisture from one part of the world to another, showers us with rain and snow, acts like a blanket to retain heat, and presents fantastic and beautiful sunrises and sunsets. The troposphere is only about 5 miles (8 kilometers) deep over the poles and up to about 12 miles (19 kilometers) in depth at the equator, but despite its small size, it is the only zone most of us will ever experience at first hand.

The temperature and density of the atmosphere within the troposphere decline with increases in altitude. It is thick and dense at sea level, and it thins out rapidly in the higher elevations to where, at about 18 miles (28 kilometers) elevation, there is virtually no air at all. You can feel this change every time you ride an elevator in a tall building or drive over a high mountain pass. As the pressure inside your body tries to equalize with the lessening air pressure outside, your ears "pop." Athletes from nations that are at low elevations experienced rather difficult problems of adjustment during the 1968 Olympic Games in Mexico City, because they had to compete in mile-high atmosphere, even though their lungs were adapted to the denser air found at near sea-level. Think of the problem of adaptation faced by the Indians who migrated to the high Andes of Peru thousands of years ago. Through the process of natural selection (survival of the fittest), they were able to adjust by increasing the size of their lungs and the density of the red blood corpuscles in their systems.

Depending on a region's temperatures, moisture, and the presence or absence of winds, the temperature of the air declines at an average rate of about 3.5°F per 1000 feet (0.6°C per 100 meters), until the upper limit of the troposphere is reached (the **tropopause**). Here, the temperature remains fairly constant at about −65°F (−54°C), and there is little or no vertical movement of winds. An interesting feature of the space occupied by the tropopause is the presence of winds that blow in a horizontal direction at tremendous speeds. These high-altitude winds (which we shall consider in greater detail later in this chapter) sometimes reach velocities of 200 miles (320 kilometers) per hour and have a strange, tubular shape. They are called *jet streams*.

Above the tropopause is a region termed the **stratosphere.** This upper level of the homosphere extends skyward to an altitude of about 50 miles (80 kilometers) over the equator and slightly lower over the poles. Air is still existent here, but it is considerably thinned out. Vertical wind movement is minimal, thus creating little in the way of air turbulence—a fact much appreciated by jet travelers and pilots.

An important feature of the stratosphere is the presence of small amounts of oxygen that are altered by the bombardment of incoming ultraviolet radiation into a heavier oxygen compound called **ozone.** Normally, an oxygen molecule consists of two atoms clinging together like Siamese twins, but the attack of solar energy causes a third atom to be added—to make it *triatomic oxygen*. As the ozone becomes agitated with further ultraviolet radiation, the atoms actually absorb these deadly rays. Without this protective barrier, called the **ozone layer,** much of life on earth would disappear within a matter of a few minutes. The only life forms that would be able to survive would be those that have their skeletons on the outside—such as cockroaches, which have survived similar forms of radiation emitted during nuclear bomb tests. When our astronauts are out "moon walking," they do not have this ozone barrier to protect them, and they would be slowly but seriously burned (like a deadly, delayed-action sunburn) if they did not wear protective suits. It is interesting to note that some meteorologists consider this ozone layer so important that they would like to term this region the *ozonosphere*.

STRUCTURE OF THE ATMOSPHERE: TEMPERATURES Another method of classifying the atmosphere is by temperatures. The atmosphere is divided into several layers, or spheres, according to *thermal* (from the Greek *therme*, "heat") characteristics (Fig. 4–3). This does not imply that the various zones or regions have absolute boundaries or have completely different characteristics. The earth has but *one* atmosphere, which happens to vary from one area to another. Aeronomists using this newer system have maintained some of the old terms and added new ones. The lowest level is still called the troposphere, and almost all of the earth's weather is contained within its sphere. The stratosphere is next, and its characteristics show how "opposite" it is to the troposphere. Whereas the temperature of the air declines in the lower region,

FIG. 4–3 Earth's Atmosphere, Classified by Air Temperatures. (After A. N. Strahler, *The Earth Sciences*, 2d ed., New York, Harper & Row, 1971.)

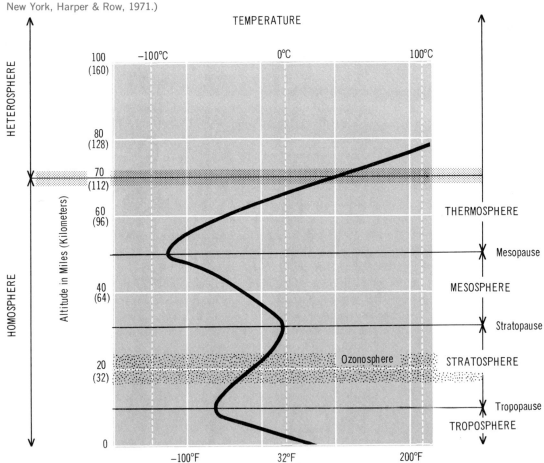

reaching a low of $-65\,°F$ ($-54\,°C$) at the tropopause, the temperature through the lower stratosphere remains near that level until an altitude of about 12 miles (20 kilometers) is attained. Then the temperature begins to rise rapidly. By the time a height of 30 miles (47 kilometers) is reached, the temperature has climbed to a level near or not far below freezing.

Beyond the upper limits of the stratosphere (the **stratopause**) another reversal takes effect, and the temperature drops rapidly as the distance from earth increases. This is the **mesosphere** (from the Greek *mesos,* "middle"), and its characteristics are still being studied. The most striking feature of the middle zone found so far is the sudden and intense decrease in temperature—from near $0\,°C$ at the stratopause down to the lowest temperatures ever recorded in our atmosphere, $-137\,°F$ ($-94\,°C$) at an altitude of 50 miles (80 kilometers). The upper limit of the mesosphere (**meso-pause**) coincides roughly with the top edge of the homosphere. To many, this is the dividing line between the upper (outer) and lower atmospheres.*

Beyond the mesopause, the temperature remains constant until an altitude of around 56 miles (90 kilometers) is reached, then once again a steady increase in temperature occurs, a process that continues on into space. This region, which includes the previously discussed ionosphere and exosphere, has been given the name **thermosphere** (and might also be called the heterosphere).

Air Temperatures

In our discussion of the various methods used to classify the structure of the earth's atmosphere, you undoubtedly have noted that in each method temperature is an important factor. You might also recall from our discussion in chapter 3, regarding the movements of the earth, that temperatures vary across the face of the earth. This will become more understandable when we have examined the seasonal distribution of air temperatures in the various latitudes.

WORLD DISTRIBUTION OF AIR TEMPERATURES As the earth revolves around the sun in its elliptical orbit, and because of its tilt from the plane of that orbit, not all portions of the earth's surface share equally in the receipt of solar energy. This tells us that the most important controls on our weather and climate are latitudinal position and the angle of the sun's rays, because the amount of insolation any area receives is dependent on these factors.

One way to clarify our understanding of the differences in temperature across the face of the earth is to make a map showing these differences. We could call such a map a *world temperature map* or an *isothermal map.* An **isotherm** is a line connecting points of equal temperature and separating regions of higher temperature

*To obtain exact figures of the altitudes of the various "spheres" that make up the structure of the earth's atmosphere, consult: *Atmosphere Supplements, 1966,* published by the Environmental Science Services Administration, National Aeronautics and Space Administration, and the U.S. Air Force, U. S. Government Printing Office, Washington, D.C. 20402.

from regions of lower temperature. At least, the isothermal boundary lines connect places where the *average* yearly temperatures are the same. Further, an isotherm is based on sea-level (or close to sea-level) temperatures, ignoring abrupt variations caused by mountains, valleys, deserts, lakes, or other features that do not cover a vast expanse. Obviously, an isothermal map of the world, being drawn to a small scale, could not indicate every change or difference in a region's temperature.

To further clarify, let us consult isothermal maps of the world. Figure 4–4 shows January isotherms, and Fig. 4–5 shows July isotherms. Note how the isotherms are drawn across the face of the earth in a general east-west direction, with the higher temperature isotherms closer to the equator than the isotherms of lower temperatures. This is proof of the importance of the interrelationship between insolation and a region's latitude. Compare the two maps, and you will see that the July isotherms migrate to the north as the Northern Hemisphere tilts toward the sun.

Temperature gradient Isothermal maps show the **temperature gradient,** which can be likened to the slope of land from high to low elevations, except that it tells us the rate of change (the slope) in air temperatures from one region to another across the earth, whether the change from warm to cold or cold to warm is abrupt (steep slope) or gradual (gentle slope). If two isotherms are close together, it means that the temperature gradient is steep, that temperatures change abruptly from one place, across the isotherms at right angles, to the other place. For example, the January isothermal map (Fig. 4–4) shows isotherms closely packed together over Central and southern Asia, yet spread far apart over equatorial Africa. This means there is a rapid change in temperatures (in a north-south direction) over Asia and a gradual change over Africa. If you will now consult the July isothermal map (Fig. 4–5), you will note how the isotherms are spread far apart over Asia, considerably further apart than in January, thus denoting a gradual change in temperature from north to south during the summer months.

Land and water distribution You may have noticed another peculiarity: The pattern of the isotherms over the continents is quite different from that over the oceans. This indicates to us that the difference in land and water distribution is another important control of the isothermal pattern. For both of the maps, check the isothermal patterns over the landmasses and the oceans. Focus on the isotherms for 30°F (-1.1°C) and 50°F (10°C) in the Southern Hemisphere and for 50°F and 70°F (21°C) in the Northern Hemisphere. You will note that the temperature gradient is lower (because the isotherms are more widely spaced) in the Southern Hemisphere than in the north. Further, the isotherms are straighter and lie more in an east-west direction. Why does this happen?

For an answer, compare the extent of the landmasses and the oceans in the two hemispheres. They are not the same. You may remember that in an earlier discussion we said that land and water both assert major controls over temperature (heating and cooling), but that the oceans tended to modify extremes, while landmasses tended to exaggerate them. Again, compare the amount of land south of latitude 20° south with the expanse of land north of latitude 20° north. When you think of that vast stretch of land in the Northern Hemisphere absorbing solar radiation over the

FIG. 4-4 Selected January Isotherms.

FIG. 4-5 Selected July Isotherms.

summer months, you know that surface temperatures will have to be high. At the same time, the great oceans of the Southern Hemisphere allow only slight changes in temperature from winter to summer.

Because of the intense heating of the continents, the isothermal pattern bends northward over North America and Eurasia during the summer months, then veers southward as the land masses cool under the grip of winter. Consider the problems such extremes cause the peoples of Bangladesh. During the summer, interior Asia heats up, creating a large area of ascending hot air. Cooler air over the Bay of Bengal and the Indian Ocean is drawn in to replace the Asian air that has moved upward. This cool (not cold) air is moisture laden, and as it rises over Bangladesh and strikes the Himalayas, condensation occurs and precipitation follows. These summer monsoons normally bring benificent rains to the parched farmlands, but sometimes they cause flooding and considerable loss of life.

Ranges in temperature   Another type of isothermal map that can aid us in understanding the differences in weather and climate across the face of the earth is a map showing the difference between summer and winter temperatures. Such a map (Fig. 4–6) would indicate which areas of the earth have the greatest ranges. Where would you expect to find the greatest range or variation: Over land? Over the Sea? In the Northern Hemisphere or the Southern Hemisphere? After you have studied this map, you will see that nowhere in the Southern Hemisphere does one find as great a temperature variation as in the Northern Hemisphere, and only in the Great Australian Desert and the Pampa of Argentina is there a variation as great as 30°F (17°C). It is a different story in the north, for in Siberia the range exceeds 110°F (61°C). How would you like living in such a place, where the temperature drops to −40°F (−40°C) in winter, then rises to 70°F (21°C) in midsummer?

Sensible air temperature   Do you know the difference between the actual air temperature (what the thermometer says) and the way your body *feels* it? This is called the **sensible temperature** and is affected by wind and humidity (moisture in the air) as well as by the temperature. Suppose you and a friend were spending the night in a well-insulated mountain cabin during the winter. You are in the living room sitting close to the fireplace; the other person is outside skiing. You decide to move into the kitchen, leaving the temperature of 70°F (21°C) for a room that is 60°F (15.6°C). Your friend comes into the kitchen from the freezing temperature outside. Will the temperature of the kitchen feel the same to both of you? No, for you will feel the coldness of the drop in temperature, while your friend will feel only the tremendous increase. You will shiver; your friend will perspire.

Air Pressures and Circulation

Air is in constant movement throughout the troposphere and all over the earth. It moves because of differences in the weight of air in different parts of the world. It moves horizontally as advective wind, vertically upwards as convectional currents, and downward from high altitudes to lower elevations as subsidence. And it moves

FIG. 4-6 Annual Range in Temperature.

Less than 5 F°(3 C°)

5 F°(3 C°)– 10 F°(6 C°)

10 F°(6 C°)– 20 F°(11 C°)

20 F°(11 C°)– 40 F°(22 C°)

40 F°(22 C°)– 60 F°(33 C°)

60 F°(33 C°)– 70 F°(39 C°)

70 F°(39 C°)– 80 F°(44 C°)

80 F°(44 C°)– 90 F°(50 C°)

90 F°(50 C°)– 100 F°(56 C°)

100 F°(56 C°)–110 F°(61 C°)

Greater than 110 F°(61 C°)

from high-pressure (heavy air) regions to low-pressure (lighter, less dense air) regions. One might say that the air of our atmosphere moves in circles, and that is why we apply the term *circulation* to the movement of air.

Air pressure means, technically, the weight of a column or mass of air. It is measured by instruments called **barometers** such as the **aneroid** barometer (a metal container that expands or contracts with decreasing or increasing air pressure) and the **mercury** barometer (which is more accurate, because it measures the weight of air by the effect of air pressure on the mercury contained in a glass tube). Standard (or normal) atmospheric pressure at sea level is equal to about 29.92 inches (74.8 centimeters) of mercury, which is also expressed as 1013 millibars.* The heavier the air, the higher the column of mercury rises in its tube, and if the pressure drops, so does the column of mercury.

WORLD DISTRIBUTION OF AIR PRESSURE SYSTEMS   Like air temperatures, air pressures are not the same everywhere on earth. These variations are caused by differences in latitude, receipt of insolation, seasons, air temperatures of landmasses and water bodies, and by the presence of forests, grasslands, and snowfields. These differences have an important effect on mankind, and just as we learned more about the effect of temperatures on climate and weather by constructing maps made up of isotherms, we can learn much about the influence of atmospheric pressure by constructing maps composed of **isobars.** An isobar is a boundary line that connects points or places having the same or equal atmospheric (or barometric) pressure and that separates areas with differing pressures. Figure 4–7 is a world isobaric chart for the month of January, and Fig. 4–8 is the chart for July. By comparing them we learn that the isobaric patterns are governed by the earth's receipt of solar energy. As we learned from our study of air temperatures, the Northern and Southern hemispheres do not receive the same amount of insolation at the same time. The

* *Millibar,* usually abbreviated *mb,* is an international unit for measuring air pressure.

Aneroid Barometers.   The barometer on the right shows the pressure of the moment; the one on the left keeps a permanent record.

FIG. 4-7 Selected January Isobars.

Low-pressure cells   High-pressure cells

Surface winds   1014   Pressures in millibars

FIG. 4-8 Selected July Isobars.

various pressure systems all migrate north and south about 15 degrees of latitude with the changing seasons.

You will also note that the different pressure regions seem to fit into something similar to zones that stretch across each of the hemispheres in a latitudinal (east-west) direction. A generalized sketch of the world's pressure belts or zones would look like Fig. 4-9 if the isobars were "averaged out," or reduced to what they would be if the localized conditions and differences were omitted from consideration. Note that there is an alternation of high- and low-pressure zones from the North Pole to the South Pole.

The key to the world picture is found in the equatorial regions. Why? Because there is no "let up" in insolation there. More solar energy is received on and near the equator than in any other latitude. As the air is heated to an extreme, its molecular activity is increased. The air masses begin to expand, becoming lighter in weight and density. As they rise, the *equatorial lows* are created. The heated air continues to rise until it reaches the tropopause, where it is forced to split into two masses: One moves north, the other south. This poleward-moving air eventually begins to cool. Some of it descends near the 30° parallels of latitude north and south. As these masses of cooling air subside, high-pressure cells are formed (the *subtropical highs*). On and near this region of high pressure can be found the earth's major desert areas. In the Northern Hemisphere are the Sahara, Arabian, Thar, Sonora (northern Mexico), and Mojave. In the Southern Hemisphere are the Atacama of Chile, the Kalihari and the Namib of southern Africa, and the Great Sandy of Australia.

The remaining masses of upper-level, poleward-moving air begin to subside over the poles. Here, the air is extremely cold, owing to the scarcity of insolation and a high albedo, and the molecules of the air slow down and begin to compact, thus forming high-pressure areas. On the earth's surface, the air moves from the poles toward the equator. At the meeting place between the cold, dense air flowing toward the equator and the warmer, poleward-moving air from the subtropical highs, the warmer air is forced aloft (since it is lighter and less dense), thus creating a low-pressure zone (*subpolar lows*). There is a difference in the subpolar lows of the two hemispheres: The Northern Hemisphere, because of its vast land areas, does not allow for one long, continuous belt of low pressure. High pressure builds up during the cold winters, because the continents get very cold and the air above compacts into high pressure. Individual low-pressure storms continually are being created in this fluid line of low pressure, and like large waves, they invade the midlatitude regions of North America and Eurasia.

Each of these pressure regions, over a period of time, has been given a popular name (we might call them nicknames). Each region has its own characteristics and peculiarities that make it stand out as being different.

The equatorial region is a place of high humidity, unstable air masses, permanently fixed low-pressure cells, and the regular occurrence of thunderstorms and water-spouts. The winds that converge within this region are variable, shifting, and generally rising winds. In bygone days, they were not to be trusted by sailing men. Ships attempting to cross the Atlantic in the tropics often found themselves wallowing helplessly for long periods of time, days without sufficient wind to move them

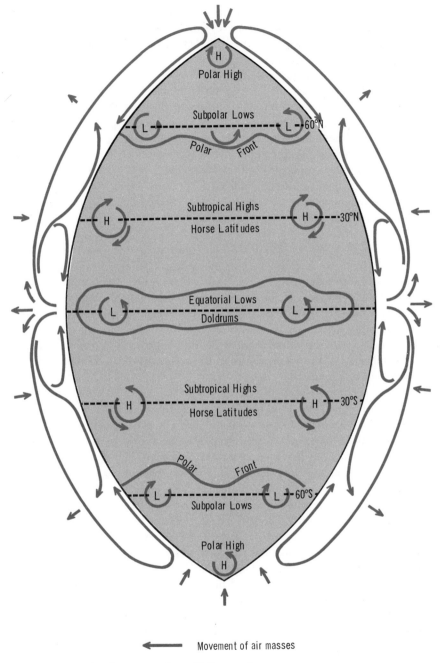

Movement of air masses

FIG. 4-9 Generalized Diagram of World Pressure Systems.

forward. Climatologists call this region the intertropical convergence zone (ITC), because this is where the tradewinds converged, but the sailors used another term: the **doldrums**. Englishmen aboard the becalmed ships became depressed at their enforced idleness and moaned that they were in the "doldrums." The nickname stuck and is now in common usage.

North of the doldrums is another region of variable, unsteady winds that often caused sailing ships much anguish. The subtropical high-pressure zone earned the name **horse latitudes**, and there are a number of opinions as to how this nickname came to be used. The most prevalent one attributes the name to the Spaniards who sailed to the New World with cargoes of horses. If becalmed for a long period of time, unable to take advantage of what winds did blow, or if the winds blew constantly toward them, making it difficult to proceed, and if the supply of feed and water was running out, the captains would "lighten ship" by throwing some of the horses over the side.

WINDS AND WIND SYSTEMS OF THE WORLD   During our discussion of the world's air pressure systems we have referred to the winds that move from one pressure region toward another, but we have not explained the reasons behind their direction and speed. Simply, winds are created out of the differences in air pressure, and they move from a high to a low because the air in the high is denser (the molecules are more compacted) and in the low, less dense. Winds do not, however, move at a constant speed, nor do they blow in a constant direction, for they are always under the influence of three basic controls: the pressure gradient, friction, and what is termed the *Coriolis effect.*

Pressure gradient   The **pressure gradient** is the determining factor in the initial speed and direction of the wind. It tells us whether the wind will be slight or moderate—or a full-blown gale. The pressure gradient indicates the change from a high-pressure to a low-pressure area and the direction of wind movement. As we have said, pressures build up to a high pressure cell (a high) in a region where the air is subsiding and the molecules are becoming packed more tightly together. As this occurs, the air must move out and away, and it does so in a clockwise direction in the Northern Hemisphere. Air rising from a heated surface creates a low-pressure cell (a low), and the air tends to spiral upward and inward in a counterclockwise direction. This inward spiraling is called *cyclonic* and the outward spiraling is *anticyclonic;* thus lows are also referred to as *cyclones* and highs as *anticyclones.* The direction of spiraling would be just the opposite for lows and highs in the Southern Hemisphere.

The pressure gradient can be found by consulting an isobaric chart, such as is shown in Fig. 4–10. On this chart, you can see a low-pressure cell bordered by two high-pressure cells. The atmospheric pressure is 1018 millibars in one high and 1015 in the other, while the low has a pressure of only 994 millibars. Note the two arrows: These are pressure gradients and are drawn from the high to the low, crossing each isobar at right angles. If the isobars are close together, the gradient's slope will be steep, the change in pressures obviously is great, and winds blowing along the gradient will be strong (high A to low B). If the isobars are far apart, the slope is gradual,

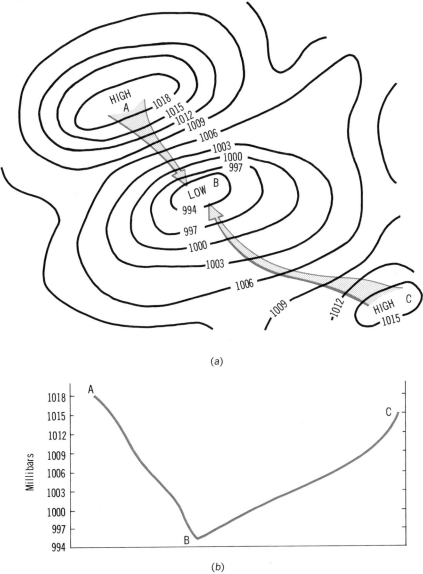

FIG. 4-10 **Pressure Gradients.** (*a*) is a rough sketch of a typical isobar chart, showing the highs and lows. (*b*) is a cross section of the isobar chart, drawn from points A to C.

as the pressure change is not abrupt, and winds, if any, will be light or moderate (high C to low B). While pressure differences start winds moving, and the pressure gradient determines the beginning speeds and directions, other controls soon take over and affect both, especially if any great distances are involved (either horizontally or vertically).

In order to further our understanding of the pressure gradient force, let us consider the winds called *land and sea breezes*. These are winds that alternate back and forth between bodies of water and the nearby land (Fig. 4–11). People who live on a coast are well aware of these local winds that blow from a lake or ocean to the land during the day and then turnabout at night and blow the other way. The reason for this phenomenon is that during the daylight hours (when insolation is at its highest) the land warms at a faster rate than does the body of water. As a result, the heated air rises over the land (thereby creating a low-pressure area), inviting the cooler air from the lake or ocean to rush onshore. At night, the land cools off rapidly (thereby creating a high-pressure area), while the ocean or lake is still warm from its accumulation of heat during the day. Now it is the warm air from the ocean that is doing the lifting, and it is the cooler air from the land that is blowing offshore.

A much larger version of these land and sea breezes is the wind system known as the **monsoon,** a common occurrence in many midlatitude regions. There are

FIG. 4–11   Land and Sea Breezes.

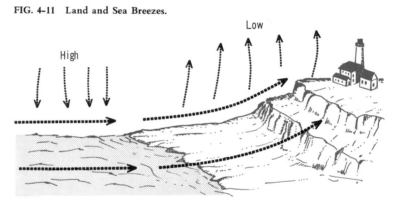

DAY Sea breezes blow onshore.

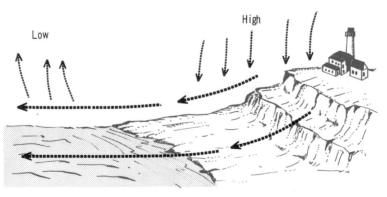

NIGHT Land breezes blow offshore.

winter monsoons and summer monsoons, and the difference between the two is in the direction they move. Winter monsoons may be likened to offshore (land) breezes, for the land is much colder in winter than are the oceans. Thus a high-pressure cell builds up over the land, and the air moves outward over the ocean, which has retained much of its warmth from the past summer. This is a time of relative stability, for the air masses are cold and dry. The summer monsoons, however, are a different matter. In both the southeastern United States and Central Asia, the land heats up rapidly, thus building a large low-pressure system, which pulls in the cooler and moist air from the nearby oceans. Figure 4-12 illustrates the invasion of a summer monsoon into Bangladesh and northern India. As we mentioned earlier, the heavy rains resulting from the onshore movement of moisture-laden winds can be both beneficial and disastrous—depending on the severity of the storm and the amount of precipitation that falls.

Friction   As winds blow along the earth's surface, a certain amount of resistance is offered by the land or the sea, and the velocity of the wind and the smoothness of its passage will be affected. The amount of **friction,** manifested by the degree of slowing or the creation of turbulence, depends on the unevenness (roughness) of the surface. As might be expected, gently undulating snowfields will not offer as much resistance as high and rugged mountains. A level plain will not impede the wind's flow as will hilly or mountainous terrain. It might be noted that friction does not affect high-level winds; it is normally restricted to altitudes of less than 2000 feet (600 meters).

Coriolis effect   We have shown how winds move from high-pressure to low-pressure areas, governed primarily by the pressure gradient, and the result should be that winds moving from a high-pressure cell move directly toward a low-pressure area. Such does not happen, however, where winds travel more than short distances, because another control asserts its influence—the **Coriolis effect.** Coriolis (named after the French scientist who first observed it) has to do with the rotation of the earth. A look at a map of the world's wind systems (Fig. 4–13) will disclose that winds that "should" move directly north or sourth do not. Rather than blow from one pressure system to another, they appear to alter their routes, their directions, almost 90 degrees from their intended path: to their *right* in the Northern Hemisphere, to the *left* in the Southern Hemisphere. Why does this happen?

Let us consider winds blowing south from the North Pole. They begin as north winds—and end up coming from the east! They are no longer "north winds," they are the polar **easterlies.*** Winds moving out of the subtropical highs and heading north appear to change direction and swing to the east. We call these winds the **westerlies,** and they are common to most of the United States. Other winds blowing out of these subtropical highs start off toward the south, but they, too, veer to the

---

* In wind terminology, the *direction* of a wind is always given in terms of its birthplace, its source. Thus, a "west wind" is one blowing from the west; an "easterly" comes from the east; a "nor'wester" is a wind blowing from the northwest.

FIG. 4-12 Summer Monsoon from the Bay of Bengal into Bangladesh and India.

right and approach the tropics from the northeast, so we call them the northeast trade winds (or northeast trades).*

The same thing is happening in the Southern Hemisphere: The winds alter their directions, but their alteration is to their left. Winds approaching the equator from the southeast are the southeast trades, then, in order, there are the westerlies and

*The word *trade* came to us from the early European traders who used the fairly constant northeast winds to move their sailing ships from Europe to the New World.

the south polar easterlies. This alteration of direction applies not only to winds, but to anything that freely moves over long distances: a migrating flock of birds, a high-flying jet, or ocean currents.

You may now ask this question: Does the wind *really* change its true direction, or does it merely *seem* to do so? For the answer, let us consult Fig. 4–14. This diagram shows the rotating earth and the deflection of its wind systems. The solid arrows indicate the direction the winds would move if the earth did not rotate, and the dotted-line arrows show the path the winds do take in relation to the earth's surface. A wind blowing from the North Pole (point A) does not reach point B, for it appears to be heading directly for point D. Why? Because the earth rotates from west to east and is moving *under* the flowing wind. Believe it or not, the wind did *not* change its direction. An observer on the ground at point D would take an oath that the wind hitting him in the face was indeed coming from the east, but to a satellite or an observer in the sky (if the wind were colored so that it were visible) the wind

FIG. 4–13 Generalized Diagram of the World's Wind Systems.

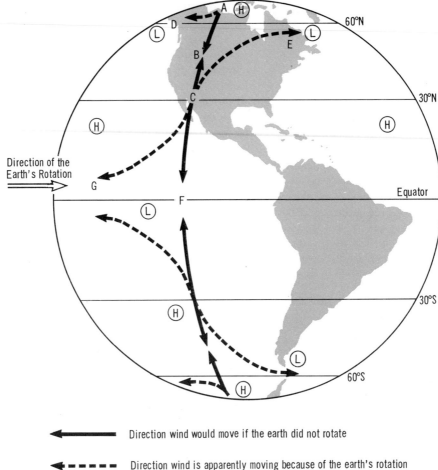

Direction of the
Earth's Rotation ⟹

Direction wind would move if the earth did not rotate

Direction wind is apparently moving because of the earth's rotation

FIG. 4-14 The Coriolis Effect.

would not have changed direction at all. It would have moved from its beginning
to the end in a true north-south direction.* The same thing will happen for a wind
blowing from point C toward point F: It will end up blowing (apparently) toward
point G. What about winds coming north from point C? If the earth is also moving
from west to east beneath these winds, why does the wind swing away from B and
veer toward E, to become the westerlies? Why does it not bend to its left, as do
the polar easterlies and the northeast trades? Well, it does not for the simple reason
that it cannot. The earth is rotating faster near the equator than it does at the higher

*In order to prove this to yourself, take a piece of cardboard and cut it into a circle. Hold a marking
pen at the center and have someone rotate the card while you concentrate on drawing a straight line
to the outside of the circle. What happens? If the cardboard is rotated in a counterclockwise direction
(to simulate the Northern Hemisphere), your line will curve to its right. If the card is rotated clockwise
(simulating the Southern Hemisphere) the line will curve to its left.

latitudes, and winds moving north from the subtropical highs are moving too fast for the earth beneath them. It is all because of the greater speed of the earth's rotation at point C than at point A.

Look at it this way: On a calm day, the air rotates with the earth as it spins, but we think of the earth and the atmosphere as being still, nonmoving. If the air begins to move over the spinning earth, the situation changes. Because the rotational speed varies with the latitude (greatest at the equator, least at the poles), we can liken this to a freeway in your city. A freeway has several lanes (like parallels of latitude), some for fast-moving and some for slow-moving traffic. When your car (wind) moves from one traffic lane to another, and you maintain your original speed, you will be going either faster or slower than the prevailing traffic in the new lane. If you were driving fast and moved into a slower traffic lane, you would immediately flash past the slower moving cars, and to them, you would be veering to their right. Using the same analogy, if you were going slow and moved into the fast lane, to the speeding cars you would appear to be moving to their left.

It should be noted that the Coriolis "force" is strongest at the poles, decreasing in influence at latitudes near the equator. Along the equator the Coriolis effect is at its weakest, and other controls take charge of the speed and direction of winds.

AIR MASSES   Everything in our atmosphere is tied in with temperature differences and the resulting buildup of air pressure regions across the face of the earth. As we have seen, low-pressure cells form over land or water bodies that have been heated by a heavy receipt of insolation, and high-pressure belts, or cells, form where the air has been cooled and is settling. While the heating and cooling processes continue, the air masses over the regions will assume their special characteristics or properties. For example, consider the northernmost portions of Canada and Siberia, lands that are very cold. Air masses forming over these regions will assume the cold and dry properties of the region, and when they move out, they will move as masses of cold, dry air. An air mass over a tropical desert, such as Africa's Sahara, will be much like the hot and dry land beneath it. Equatorial waters not only heat the air above, but will also send moisture into the air; thus the resulting air mass will be warm and moist. So, in order to have a greater understanding of the movement of winds and the varying results of their arrival, we must consider their points of origin, the source regions that gave them birth.

Source regions   As a beginning, let us consult Fig. 4–15, which illustrates (in a generalized way) the **source regions** for the different air masses. Meteorologists have devised a code to simplify matters. The capital letters (A, AA, P, T, E) refer, roughly, to latitudinal position, and the lower case letters (c, m) apply to the actual earth surface, whether land or ocean. The letter A stands for *arctic;* AA for *Antarctic;* mP, *polar maritime,* for source regions, such as the waters off the Aleutian Islands or near Greenland; cP, *polar continental,* for regions such as the cold interior of Canada or Asia; mT, *tropical maritime,* for tropical waters, such as the Caribbean, the Gulf of Mexico, or the mid-Atlantic: cT, *tropical continental,* for air masses forming over the hot deserts, grasslands, or forests of large land areas within the equatorial region.

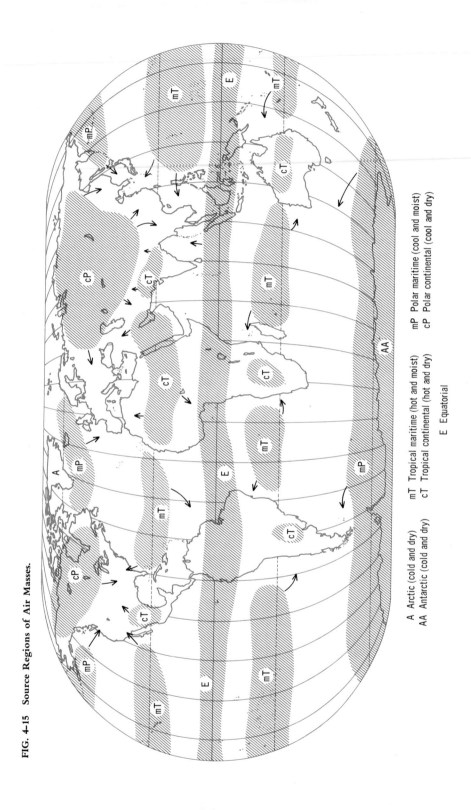

FIG. 4-15 Source Regions of Air Masses.

A Arctic (cold and dry)   mT Tropical maritime (hot and moist)   mP Polar maritime (cool and moist)
AA Antarctic (cold and dry)   cT Tropical continental (hot and dry)   cP Polar continental (cool and dry)

                                 E Equatorial

118

E, *equatorial,* is a relatively new addition as a source region and is confined to the area between 10° north and 10° south.

We have previously discussed many of the winds resulting when air masses of different properties collide or, at least, when one of them intrudes on another's territory. As an example, the monsoons that periodically sweep over India and Bangladesh are the result of a very hot polar continental air mass building up over interior Asia during the summer, thus inviting the intrusion of slightly cooler tropical maritime air from the Bay of Bengal and the Indian Ocean. The reverse occurs in winter, when the Asian polar continental air mass is cold and dry and sweeps down from Tibet's high plateau and over the Himalayas onto the lands surrounding the Bay of Bengal.

Now let us consider an important and very interesting outgrowth of the meeting of differing air masses, as when cool air from maritime or continental polar regions of the high latitudes and warm air marching poleward from the tropical maritime or continental regions converge with air masses moving out of the subtropical high-pressure belts. Wind systems are created that are called jet streams.

Jet streams   Many strange phenomena occur in our atmosphere, but one of the least understood is the phenomenon called the **jet streams**.* These are high-altitude winds that encircle the earth at elevations of from 4 to 10 miles (6 to 16 kilometers), moving at speeds up to 300 miles (480 kilometers) per hour. Both the Northern and Southern hemispheres have at least three of these "swiftly flowing rivers of air": near the tropics, near the subtropics, and along the polar fronts. Although these winds have probably existed since the beginnings of the atmosphere, the world did not become aware of them until the 1920s. The first inkling of their presence came when weather balloons were sent aloft through gentle breezes to probe the upper reaches of the troposphere. Imagine the consternation of meteorologists, when a balloon, traveling at the rate of 15 miles (24 kilometers) per hour, and expected to take about two weeks to drift from Des Moines, Iowa, to New York City, landed in New York only four days later? The final proof that these racing winds did actually exist came during World War II, when American bombers tried to raid Japan. In order to avoid detection, they climbed into the tropopause and headed for Japan— and discovered they were not traveling at a speed of 450 miles (720 kilometers) per hour, as their speed indicators showed, but only at 250 miles (400 kilometers) per hour. They were heading directly into the subpolar jet stream.

Today, aircraft flying to Europe from North America make use of this jet stream to save fuel and time. If they are heading westward, however, it is another story. Planes returning to North America try to avoid the jet streams by flying north or south of the jet stream's errant path.

What are the jet streams? For an answer, let us consider the jet that affects North America (Fig. 4–16). This jet stream appears to follow the tortuous meanderings of the leading edge of the polar front. The sometimes violent winds are given birth

* For a thorough and interesting explanation of the jet streams, see: Elmar R. Reiter (1967), *Jet Streams* (Garden City, N.Y.: Doubleday, Anchor Books).

by the collision of cold polar maritime and continental air pushing down from the Arctic and Canada against warmer air moving north from the subtropical high-pressure belt (roughly 30° north). Turbulence is created over the upper level of the air masses, and strong winds begin to flow down the pressure gradient. As the same thing is happening over northern Europe and Asia, the result is a ribbon of rapidly moving air that encircles the globe. It is surrounded on all sides by much

FIG. 4-16 Jet Streams in the Northern Hemisphere.

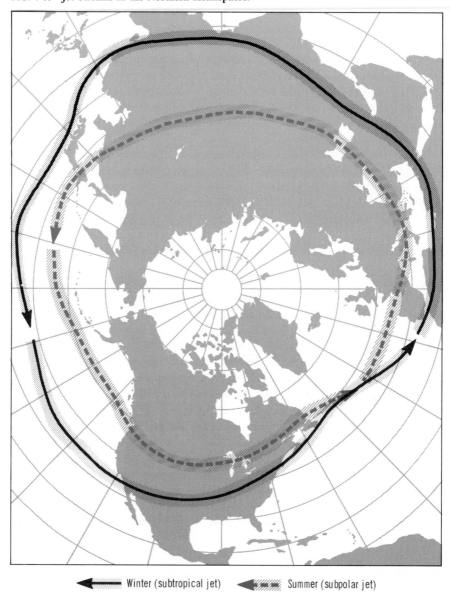

Winter (subtropical jet)   Summer (subpolar jet)

Jet Stream. These long cirrus clouds, occurring about 100 miles (160 kilometers) northwest of Dakar, Senegal, probably marked the location of a subtropical jet stream. They were photographed by a camera aboard unmanned *Apollo 6* in early 1971.

calmer air, and some meteorologists liken it to a flattened tube or hose that is about 1 mile (1.6 kilometer) in thickness and up to 100 miles (160 kilometers) in width. Within the "tube," wind speeds vary from 30 to 300 miles (48 to 480 kilometers) per hour, while wind speeds outside may be as low as 10 miles (16 kilometers) per hour, depending on the season and the geographical region. It appears that the greatest velocities occur off the eastern edge of Asia and near Greenland. It should also be noted that the jet stream is not one long continuous ribbon; rather, it is broken into sections, some short and some stretched out for over a thousand miles.

As for jet stream terminology, the summer-created jet is usually called the *subpolar jet,* and the one formed during the winter is the *subtropical jet.* Both are primarily westerly winds and exert a tremendous influence on our weather. It is believed that these jets propel storms into the midlatitudes (especially tornadoes) during the winter and early spring. They also influence summertime weather, as when the snakelike wanderings of the jet stream isolate masses of cold or warm air, causing the particular region affected to suffer an unusual cold spell or an unseasonably hot and dry period. This is probably an important spin-off from the phenomena of jet streams: They cause a mixing of cold polar air and warm tropical air.

Review and Discussion

1. Explain the difference between weather and climate.

2. Define these terms: *ionosphere, exosphere, aeronomist.*

3. What makes the sky blue? Why is it bluest overhead?

4. Explain the composition of the atmosphere by its gaseous composition. Explain these terms: *homosphere, heterosphere.*

5. Explain the difference between the troposphere and the stratosphere.

6. Why do your ears "pop" at high altitudes?

7. What is ozone? What is the importance of the ozone layer?

8. Explain the structure of the atmosphere by temperatures.

9. Explain these terms: *mesosphere, mesopause, thermosphere, thermopause.*

10. What is an isotherm? Is there a difference between the pattern of isotherms between January and July?

11. What is temperature gradient?

12. Why is the pattern of isotherms different over landmasses and the oceans?

13. Why do India and Bangladesh experience summer monsoons?

14. What is sensible air temperature?

15. Explain the development of different air pressure regions.

16. What is the difference between an aneroid and a mercury barometer?

17. What is meant by the term *millibar?*

18. What is the effect of the changing seasons on the January and July patterns of isobars?

19. What is the difference between the zones called the doldrums and the horse latitudes?

20. Explain these controls over the wind systems of the world: pressure gradient, friction, Coriolis effect.

21. What is the ITC?

22. Explain the local wind system known as land and sea breezes.

23. Why do we call a wind blowing from the west a westerly wind, rather than an easterly?

24. Why is knowledge of air mass source regions important for a meteorologist?

25. Explain these symbols: mP, cP, mT, cT, A, AA, E.

26. Explain the origin, shape, and importance of jet streams.

Additional Reading

Blair, Thomas A., and Robert C. Fite (1965). *Weather Elements*. 5th ed. Englewood Cliffs, N.J.: Prentice-Hall. A text in elementary meteorology.

Dempsey, M. (1965). *The Ocean of Air*. London: Purnell and Sons. Distributed in the United States by Ginn and Company, Boston.

Donn, William L. (1965). *Meteorology*. 3d ed. New York: McGraw-Hill.

Huschke, Ralph E., ed. (1959). *Glossary of Meteorology*. Boston, Mass.: Meteorological Society.

Lehr, P. E., and R. W. Burnett (1965). *Weather*. New York: Golden Press.

Riehl, Herbert (1972). *Introduction to the Atmosphere,* New York: McGraw-Hill.

Starr, V. P. (December 1956). "The General Circulation of the Atmosphere," *Scientific American* (offprint no. 841).

5

Moisture: Clouds and Storms

All the waters run into the sea;
yet the sea is not full;
unto the place whence the rivers come,
thither they return again.

ECCLESIASTES 1:7

Our investigation into the earth's dynamic atmosphere has shown how important water is in contributing to and modifying weather and climate. Without moisture, our earth undoubtedly would be uninhabitable, much as the planets Mercury or Mars. Fortunately, however, our earth has sufficient gravitational attraction to hold the atmosphere in place—as it has for more than 2 billion years. Further, the earth is situated at the right distance from the sun to allow moisture to exist in all of its three forms: vapor, liquid, and solid. There are a number of theories as to how the earth received its water supply, but the one most widely accepted suggests that the earth was once a fiery liquid that cooled slowly through countless millions of years. In the beginning, the surface was too hot to allow water molecules to remain, and most of them immediately boiled away into the atmosphere. As time passed, the crust cooled sufficiently to allow the falling water to accumulate in great pools.*

At first, there may have been only one ocean that covered the earth's rock mass, but, in time, the earth's crust began to break up, and the landmasses rose above the surface of the seas. As the ages passed, solar radiation continued to strike the waters of the oceans, causing some of the moisture to evaporate and ascend to the skies, where it cooled, condensed, and was precipitated. Thus, a process that we might call the "cycle that permits life" was born. Hydrologists call this the *hydrologic cycle*.

The Hydrologic Cycle

The word *cycle* can be defined as "a repeating or recurrent series; a succession of events that repeat themselves in an orderly manner over a long period of time." The term *hydrologic* means "pertaining to water." We can, therefore, combine these two words and say that the **hydrologic cycle** is the "water cycle," which is another way of saying that water is involved in a recurring series of events during which it changes form in an orderly manner.

Evaporation is an important step in the hydrologic cycle (Fig. 5–1), the result of the transformation of water from either a liquid (water) or a solid (ice or snow) state to the gaseous (water vapor) state. The water vapor is carried upward by convectional currents and is spread throughout the air masses that lie over the water body. Onshore winds then push the moisture-laden air inland. Condensation occurs through orographic lifting, convection, or frontal uplift in cyclonic storms, and

*For those who would like to pursue this topic further, see: L. B. Leopold and K. S. Davis (1970). "A Sun-Powered Cycle," *Water* (New York: Time-Life Books), chap. 2. The authors are hydrologists, and their book is well written, authentic, and interesting.

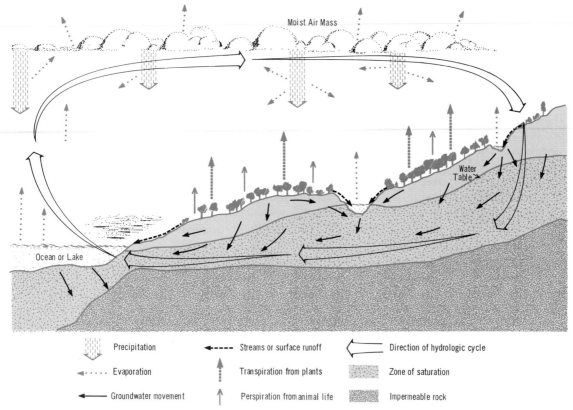

Moist Air Mass

Water Table

Ocean or Lake

	Precipitation		Streams or surface runoff		Direction of hydrologic cycle
	Evaporation		Transpiration from plants		Zone of saturation
	Groundwater movement		Perspiration from animal life		Impermeable rock

FIG. 5-1 The Hydrologic Cycle.

precipitation falls to the surface of the earth. Some moisture falls as snow or hail and is held at the higher elevations (or high latitudes) as a sort of "solid reservoir" until it melts. Thawing and melting add water to the underground supply.

Most precipitation is in the form of rain, and much of that becomes surface runoff, flowing through streams to larger streams, rivers, lakes, and, eventually, the oceans. Some water infiltrates downward through porous materials into what is called the zone of saturation, from where it moves by gravity flow to the ocean. Evaporation goes on throughout the entire cycle, and the water, whether in vapor or liquid form, is in constant motion.

Earth's Water Balance

How a region fares with regard to its water supply depends on its water balance. Essentially, **water balance** (or water budget) is the ratio between the receipt and the loss of moisture. It takes into consideration not only the additions through the various forms of precipitation, but the amount retained by the soil, underground reservoirs, vegetation, lakes, streams, and rivers. These totals are then related to the losses by gravity out of the groundwater zone, by evaporation into the air, and by **transpiration** (moisture emitted through leaf tissues) from plants. If the precipitation exceeds the losses, an area is said to have a favorable water balance. If the balance is unfavorable, agriculture may be severely limited, as residents of many areas of North Africa, Australia, Central Asia, and the American Southwest can testify. There are a number of contributing factors, such as the region's air temperatures throughout the year, the velocity and direction of winds, the types of vegetative cover and soils, the topography—whether smooth or rugged— population numbers and distribution, the extent of urbanization, and, most importantly, the region's location in relation to the distribution of the earth's waters.

Moisture and Precipitation

Moisture in the form of water vapor is an important part of the troposphere. It is the direct result of the change in form from a liquid (through a process called **evaporation**) or from a solid (through a process called **sublimation**, when heat causes water molecules to separate from their compacted state and rise as vapor without passing through the liquid state). Even in small quantities, water vapor is important. It supplies the earth with moisture in the form of clouds and precipitation, and it absorbs heat that radiates from both the sun and the earth.

The amount of water vapor affects the rate of water loss through evaporation from the earth's water bodies. This rate varies from rapid to slow, depending on the amount of moisture available for evaporation in a given area, the size or extent of the surface area, the salinity (saltiness) of the water, the air pressures, the presence or absence of winds, and the temperature of the air mass. These factors form a kind of "chain," and if one is changed, the entire situation is altered. For example, if there is no wind, evaporation will be less, and water might lie in ponds, lakes, or oceans for a long time. If the air temperature drops too low, the water might freeze, and there would be less evaporation. And, while there is considerable potential for evaporation from any large water body because the supply is there, the rate of evaporation will be greater over tropical waters, because of their consistently higher temperatures, than over polar waters. Ideal conditions for a high rate of evaporation would be a hot, dry, and windy day in an area plentifully supplied with moisture.

RELATIVE HUMIDITY   The term **relative humidity** is used as the standard of measurement for the amount of moisture a given air mass can hold, and it is expressed as a percentage. *Humidity* can mean dampness, moisture, or the amount of water vapor in the air. More technically, however, we can say that humidity is the amount of water vapor in a given air mass at a given temperature, as compared to the amount of moisture the air mass *could* hold if it were fully saturated and there were no accompanying change in temperature. If the daily weather report gives the humidity of your locale as 60 percent, it means that the air mass contains only 60 percent of the moisture it could hold at its present temperature. What happens if there is nearby a large body of water that sends moist air into your locale? As the moisture penetrates the air, the humidity rises. If there is no temperature change, the humidity will continue to rise until the air mass is fully saturated, or soaked, with water. If dew point is reached, condensation can occur.

This tells us that a given body of air can hold and retain just so much moisture at a certain temperature, but if the temperature is altered, so is the moisture-holding capacity of the air mass. If the temperature of the air is raised through terrestrial radiation, molecular speed within the mass is increased, and the air can hold more water vapor. Winds can cause an air mass to move, thus enabling it to expand and increase its moisture-holding capacity. A decrease in temperature on a calm, windless day will cause a drop in humidity and, thus, lower the saturation point. Dew point can be attained sooner, and some form of condensation may occur, such as fog, clouds, or ice crystals, and if the saturation point is exceeded, some form of precipitation will fall, such as hail, snow, or rain.

It is interesting to note that of all the moisture contained in our atmosphere, more than half lies below the 2-mile (3-kilometer) level.

FORMS OF CONDENSATION   As we have seen, **condensation** will occur with either one of two factors: (1) if there is an increase in the amount of water vapor in an air mass to its saturation point, or (2) if the air temperature falls to a level where its capacity to hold moisture is reduced. If this **dew point** is reached when the air temperature is at or below freezing, clouds made up of ice crystals will appear. If the dew point is reached at a higher temperature, fog or clouds made up of water droplets will form.

Fog   At sea level or in high inland valleys, fog may occur. It appears when a warm, moist air mass comes in contact with a layer of colder air close to the ground. A common fog for inland areas is **radiation fog**, called **ground-inversion fog**, which occurs when the land is chilled during the night through the loss of heat by radiation and the warmer air above has its bottom layers cooled to dew point. Such fogs usually disappear by midday or as soon as the sun's rays can warm the ground and end the inversion.

A fog common to San Francisco is called **advection-radiation fog.** This type of fog results when a warm, moist air mass (in this case, pushed by westerly winds from the Pacific Ocean) crosses a cool or cold landmass or a cool ocean current (such as the California current that moves past the Golden Gate), where it is chilled to the point of condensation. Because the warm air mass continues to converge with the cool body of land or water, the thick blanket of fog is constantly being resupplied. In addition, the fog's thickness prevents the sun's rays from warming the land surface. The fog layer, therefore, remains for much longer periods of time than the thinner ground-inversion fog.

Clouds   Probably the most important result of condensation is the phenomenon we call clouds, for clouds can furnish precipitation in the form of snow, sleet, hail, or rain. If there is a rapid upward movement of warmed air, an unstable condition will be created. Clouds will form, and if the air mass is deep enough and contains much moisture, water droplets (formed around minute dust particles) will come together to form larger and heavier droplets that fall as rain or snow.

Clouds may be classified according to their general form and the altitude at which they appear (Fig. 5–2). We use the term **stratus** to mean any kind of cloud formation that spreads out in layers over a wide area, seeming to cover the sky from horizon to horizon. Stratus clouds can be found at different elevations, but they usually form within a few thousand feet of the earth's surface. When someone speaks of an "overcast sky," he is referring to decks of stratus clouds that block out the sun. **Cumulus** clouds appear like huge puffs of cotton (or giant gobs of whipped cream). Their flat bottoms seem to rest close to the earth, while their heads tower thousands of feet into the sky. Smaller versions of this type of clouds often look like small cotton balls scattered across the sky. **Cirrus** clouds form at high altitudes, stretching out in thin, lacy, or streamerlike wisps near the tropopause.

Both stratus and cirrus clouds result from condensation occurring without any up or down movement, while cumulus clouds are the result of convectional uplift. Depending on the altitude of formation and the process, these terms can be combined to explain different or unusual clouds, such as **stratocumulus** (broad layer of thick and puffy cumulus clouds), **altostratus** (a layer of clouds at elevations of thousands of feet), **cirrostratus** (a high, but thin layer of stratus clouds), or **cumulonimbus** (from the term *nimbus,* which means "rain-bearing"). It is interesting to note that the most common type of cloud the world over is the stratus, especially from about the 30° parallel northward.

PRECIPITATION   Before precipitation can occur, clouds must form. It must be noted, however, that not all clouds yield rain, hail, or snow. Condensation and the resulting formation of clouds is the first step, but before there can be precipitation, the water droplets must combine and become sufficiently heavy in order to fall to the earth.

What forms will precipitation take? If the temperature is below freezing, snow may form as tiny flakes or hail may result from the freezing of coat after coat of moisture around the beginning nucleus of dust particles or ice crystals. If sufficient water droplets coalesce to form a raindrop and the temperature is above freezing, rain will fall.

There are three primary causes of precipitation, and all have to do with the lifting of warm, moist air masses to a zone of cooler air temperatures, where dew point can be reached and exceeded. The three types of precipitation are orographic, convectional, and frontal (Fig. 5–3).

Orographic precipitation results when a warm, moist air mass moves onshore and is forced to rise, owing to an obstruction, such as a higher, elevated landmass (mountains, upsloping plains, or plateaus). This forced change in altitude causes the air mass to lose its heat through the adiabatic cooling process.

Convectional precipitation results when a mass of unstable air rises rapidly from an area that has been heated to extremes, such as on a South Pacific atoll, in a tropical or subtropical desert, or around a warm ocean current. Such an updraft is normally quite violent in its movements as the heated air rushes upwards and cools at a higher elevation. The turbulence causes the air mass to react violently, and a cumulus cloud

Cirrus

Altocumulus

Nimbus

Cumulus

FIG. 5-2 Forms of Condensation: Clouds and Fog.

Cirrus

Altocumulus

Nimbus

Cumulus

Stratocumulus

Stratus

Fog

Stratocumulus

Stratus

Fog

132

FIG. 5-3   Causes of Precipitation.

OROGRAPHIC

CONVECTIONAL

FRONTAL

┅┅┅▶ Movement of air masses

The Windward Coast of Oahu.

begins to form—a furiously boiling white cloud thrusting up into the sky. Precipitation is short-lived, although sometimes quite heavy, and sometimes thunderstorms are created.

Frontal precipitation (also referred to as cyclonic) is the result of a confrontation between two large air masses, one warm, the other cool. If the cool air is the aggressor, the result is a cold front; if the warm air is doing the moving, a warm front develops. Consult Fig. 5–4, which presents the two types of fronts that cause precipitation. During the winter, the most likely occurrence is for a cool mass of air to move southward out of the polar front. When the cool air collides with the stationary mass of warm air, it shoves under the warm air, much like the sloping blade of a bulldozer. This forces the warm air aloft, followed by the familiar adiabatic cooling process, with condensation and precipitation resulting.

Fronts are common (and quite typical) during the late fall and winters along the Great Plains of the United States. If a warm tropical air mass is the aggressor and moves into territory already occupied by a stationary and colder air mass, the warm air front will be forced to override the cold air mass, until dew point is reached, condensation occurs, and precipitation results. This is typical of air masses leaving the Caribbean and the Gulf of Mexico to swing over Texas and the southeastern portion of the United States, there to collide with cold air masses from the north.

Figure 5–5 explains the development of a cyclonic storm along a weather front. Here, you see the counterclockwise spiraling of the low-pressure cell that converges on a warm air mass. When a warm air mass is forced aloft, completely cut off from its source and the ground, the result is an **occluded front.** You can tell in advance which front is arriving by the temperature change that occurs. If a cold front passes when the air around you is warm, the temperature will drop. The reverse occurs with the passage of a warm front. Precipitation during a warm front will be widespread and heavy, and it may last for hours or, sometimes, for days. Cold-front precipitation, on the other hand, is usually short-lived and relatively light.

We might say that orographic precipitation will be most common in regions where barriers (such as mountains) obstruct the free movement of air masses; convectional precipitation will be the normal type for tropical regions, because of the excessive

FIG. 5–4 **Frontal Activity in the Midlatitudes.**

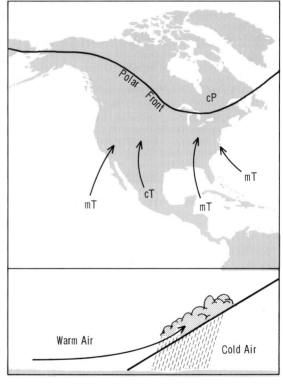

WINTER Cold fronts dominate, and the cold air mass is the aggressor.

SUMMER Warm fronts dominate, and the warm air mass is the aggressor.

(a)

(b)

(c)

insolation received; and frontal activity will be the customary precipitation for the midlatitude regions that are dominated by polar fronts.

Because rainfall is so important to mankind, it is necessary for us to know where and how much rain falls, to enable us to plan for the future by understanding how to preserve our present water supply. To do this, meteorologists collect precipitation data and then construct maps to show the areas of world rainfall and all other forms of precipitation. By connecting points of equal precipitation with lines (called isohyets), we can establish boundaries of the different regions and, thus, establish a map showing the world's average annual rainfall (Fig. 5–6). On consulting this map, it becomes obvious that the equatorial regions receive the heaviest annual rainfall; many places in the tropics receive more than 80 inches (200 centimeters) per year. Note also that areas separated from large bodies of water, or those on the lee (or rain shadow) side of mountain ranges get little precipitation. Areas that lie in the path of the trade winds, typically the Hawaiian Islands, get the heaviest rainfall. On the windward coasts of Hawaii, Maui, Oahu, and Kauai, heavy precipitation results in tropical rainforest types of vegetation, while on the downwind or lee side, the vegetative cover is much like that of Arizona or New Mexico. The same applies to the lee and windward sides of South America's Andes. Regions in the midlatitudes get little of what might be termed consistent rainfall, owing primarily to conditions created by the belt of subtropical high pressure, with its descending, compressing air.

Now, consider Fig. 5–7, which shows the average annual rainfall of Canada and the United States. The southeastern United States receives about 40 inches (100 centimeters) of rainfall, while most of the remaining states and Canada receive less than 20 inches (50 centimeters). One amazing contradiction may be seen in the coastline from the state of Washington north through British Columbia and Alaska, where 80 inches (200 centimeters) of rain falls annually. Why does this region receive so much precipitation? This is primarily because of the series of wavelike onslaughts of low-pressure cells coming from the Aleutian Islands—moisture-laden winds striking the mountain ranges, resulting in orographic precipitation.

Storms

Many storms are the result of a collision between two advancing air masses, one cold, the other warm. If you recall our discussion on pressure gradients and on precipitation due to frontal activity, you probably realize that this type of storm is the result of cyclones and anticyclones coming together. For an example, consult Figs. 5–8 and 5–9, which are weather maps for two days during the month of

FIG. 5–5 Development of a Cyclonic Storm along a Weather Front. (a) shows the aggressive advance of a cold air mass on a stationary warm air mass. (b) shows how the counterclockwise spiraling cold air has trapped the warm air and forced it aloft, thus creating an occluded front (c), with resulting condensation (clouds) and precipitation.

FIG. 5-6 Annual World Rainfall and Tropical Cyclones of the World.

Over 80 in. (200 cm)

60–80 in. (150–200 cm)

20–60 in. (50–150 cm)

10–20 in. (25–50 cm)

Under 10 in. (25 cm)

Source regions and direction of movement of tropical cyclones (hurricanes)

60°N
50°N
40°N
30°N

180° 160° 140° 120° 100° 80° 60° 40° 20°

0 400 800 miles

80 in. (200 cm)   20 in. (50 cm)
40 in. (100 cm)   10 in. (25 cm)
Under 10 in. (25 cm)

FIG. 5-7 Average Annual Rainfall: United States and Canada.

November, 1972. Inhabitants of the American Northwest, the Great Plains, and the Great Lakes, as well as, it is important to note, the citrus growers of Florida have long been aware of storms brought to them through cyclonic activity, the invasion of polar air masses. During World War II, a Norwegian meteorologist, Vilhelm Bjerknes, examined the data showing the continual onslaught of these cool lows, and it seemed to him that it was much like a battleground during a war, because frontal storms advanced in a series of waves, much like regiments of soldiers. His

FIG. 5-8 National Weather Map for Saturday, November 11, 1972. (Prepared in cooperation with the National Weather Service.)

Nov. 11, 1972

●——— Cold front

▼——— Warm front

Rain

Snow

⟵ Direction of wind

FIG. 5-9 National Weather Map for Sunday, November 12, 1972. (Prepared in cooperation with the National Weather Service.)

Nov. 12, 1972

━━━━ Cold front   ━━━━ Warm front

▨▨▨ Rain   ⋰⋰⋰ Snow

→ Direction of wind

0   200   400miles

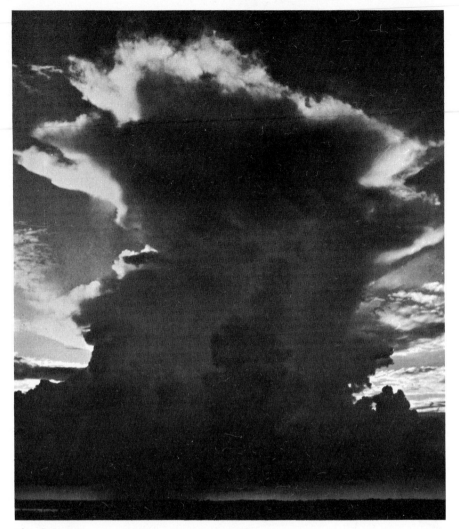

Thunderstorm!

"wave theory of cyclones" came to be known as the **polar front theory** and is still accepted today. Many of the storms that strike the North American continent, such as thunderstorms, tornadoes, and hurricanes, are the result of cyclonic action.

THUNDERSTORMS Thunderstorms are born out of the upward rush of heated air to high altitudes, which is created out of the differences in air pressure brought on by unequal heating of areas of land and sea. In the American Midwest, a region will be heated until the air begins to rise and outside air moves in; the outside air, then, is also heated and begins to rise. The low-pressure region sends its air spiraling upward in a continuous cyclonic flow (counterclockwise) to great heights, aided by what has been termed the "chimney effect," where the air at these high altitudes is expanding and moving at faster speeds, thus drawing more air upward from the ground. As condensation occurs, latent heat is released as the towering cumulus cloud is formed, and this increases the violence of the boiling cloud. As the cloud increases its height, sometimes as high as 7 miles (11 kilometers), and as temperatures within the cloud decline, rain and snow form and begin to fall, and the top of the cloud spreads out in wispy streamers.

The cloud becomes a thunderstorm when the precipitation creates a downdraft within the cloud as the subsiding air drags air behind it. Winds are violent within the thunderhead, as warm air races past cold air and the electrical field within the cloud becomes highly charged: Lightning flashes, and the sound of thunder is heard across the land. Cold air pours out of the cloud's base, accompanied by heavy rain and high winds and, sometimes, tornadoes. Eventually, the downdraft destroys the thunderstorm, for it has cut off the supply of warm air that was the impetus for the storm's beginning. Without a continuous supply of warm, rising air and the resulting precipitation, the thunderstorm loses its vitality and its heat energy and slowly settles into oblivion.

Thunderstorms normally result from the interaction between cold and warm fronts, as happens over most of the United States and southern Canada, and from the differences between the surface temperatures of land and water, as is the case in storms that occur throughout the southeastern United States, with warm cyclones (hurricanes) providing the impetus.

TORNADOES **Tornadoes** are violent windstorms that are "spinoffs" from thunderstorms, and although they are relatively small in size and short-lived, they can be very destructive.* In the beginning, the towering thunderhead is dark and foreboding, with a heavy curtain of rain descending to the ground, and it is difficult to distinguish the separation line of cloud and rain. Then, from the bottom of the cloud a funnel appears, circling wildly in a counterclockwise motion, and reaches for the earth. It may appear and disappear once or twice, then, finally, the funnel touches down and the destruction begins.

The storm sometimes moves very slowly, almost seeming motionless, and sometimes at speeds of up to 70 miles (112 kilometers) per hour. One thunderstorm may spawn several tornadoes, forming at intervals, as it sweeps across the land, and not all of them reach the ground. The path of a tornado is usually quite narrow in width, perhaps 0.25 mile (0.4 kilometer) wide, although some attain lengths of from 16 to 300 miles (26 to 480 kilometers).

It is interesting to note that the phenomena we know as tornadoes are pretty well limited to the United States. Figure 5–10 details the number of these storms that have struck the United States for a 16-year period (1953–1969). The state of Rhode Island appears to be the only one to have never experienced a tornado, and it would

---

*For informative and interesting information on both thunderstorms and tornadoes, see the National Oceanic and Atmospheric Administration's publications: *Thunderstorms,* NOAA/PA 70014 (1973), and *Tornado,* NOAA/PI 70007 (1970). They can be obtained from the Superintendent of Documents, U.S. Government Printing Office, Washington, D.C. 20402.

Tornado!

The Aftermath.

also appear that if a "tornado alley" does exist, it would include Texas, Oklahoma, Kansas, Nebraska, Iowa, Missouri, Illinois, and Indiana. Florida and the Southeast also have their share of tropical cyclones-turned-tornadoes.

The worst tornado on record "ripped across Missouri, lower Illinois, and Indiana in three hours on March 18, 1925, and killed 689 people."[1] Since "the early 1950's, the tornado death toll has averaged about 120 per year" in the United States.[2] This does not include, of course, the destruction of property that, during this century alone, must amount to billions of dollars.

HURRICANES AND TROPICAL CYCLONES We have concentrated on storms of the midlatitudes, primarily because of our concern with the United States and Canada, because they lie in the path of invading polar-front air masses. There is, however, another type of cyclonic storm, the **tropical cyclone,** which originates in the tropics. Depending on which part of the world you are in, the names will vary: *hurricane* (coming to us from the Indians of the Caribbean), *typhoon* (from east Asia and the China Sea), *cyclone* (from the Indian Ocean), and *baguios* (from the Philippines). These storms result from the rapid buildup of heated air over the warm equatorial waters and, occasionally, over the broad stretches of land along the equator. As the heated air rises, it develops a counterclockwise spiraling vortex upward, similar to the motion of a thunderstorm or tornado. When the air mass begins to move, speeds of the whirling winds may reach more than 100 miles (160 kilometers) per hour

FIG. 5-10 Tornadoes in the United States—1953-1969. The numbers indicate to total number of tornadoes for each state during the 16-year period. (U.S. Department of Commerce Bulletin, "Tornado," NOAA/PI 70007, National Oceanic and Atmospheric Administration, 1970.)

Satellite View of Hurricane Agnes.   This photo shows hurricane Agnes heading toward the west coast of Florida from the Gulf of Mexico. It hit Florida on June 19, 1972, and went on a 10-day rampage up the Eastern Seaboard, unleashing the most extensive floods in the nation's history.

within the storm, and the storm itself may advance across the open sea at a speed of from 20 to 40 miles (32 to 64 kilometers) per hour. Accompanying the violent winds will be thunderstorms and heavy precipitation.

During the summer, the southeastern coast of the United States must occasionally face these storms, as they sweep up from the tropics and move across the Caribbean and the Gulf Coast. Individual hurricanes have severely punished the land all the way from Texas up to New England, averaging about six per year. Some hurricanes also originate along the western side of Mexico and Central America, but they seldom have the strength to move far enough north to bother California.

An interesting feature of a hurricane is what meteorologists call the "eye of the hurricane" (Fig. 5–11). This eye results from the rapid spiraling of the air mass as it travels across the waters, and here, the air is calm and flocks of birds may be

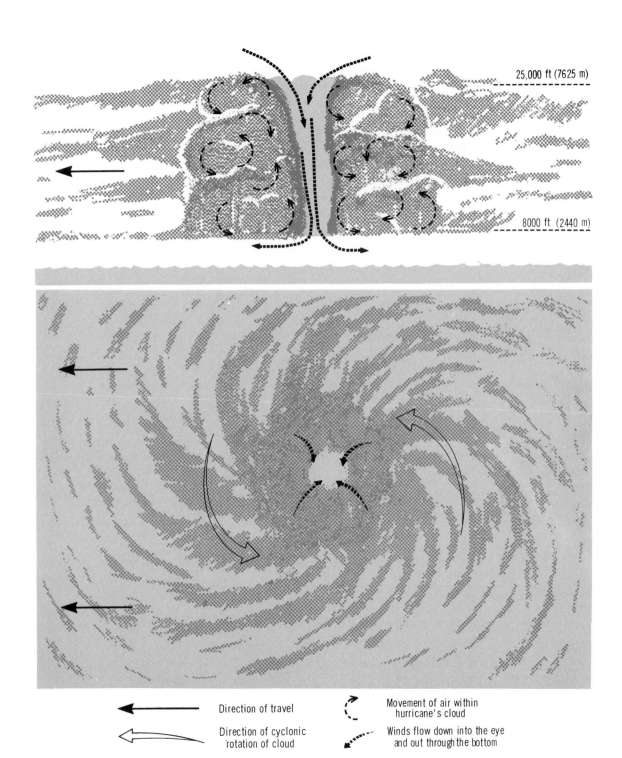

25,000 ft (7625 m)

8000 ft (2440 m)

⬅ Direction of travel	⤺ Movement of air within hurricane's cloud
⬅ Direction of cyclonic rotation of cloud	⤺ Winds flow down into the eye and out through the bottom

FIG. 5-11 The Formation of a Hurricane.

found flying serenely around in circles, not daring to brave the angry winds all around, waiting patiently for the storm to die. The eye gives a few moment respite for a storm-tossed ship or a wind-battered city, but then comes the other side of the storm and winds coming from the opposite direction, and the storm-ravaged area goes under seige once more.

Worldwide, tropical cyclonic storms seem to have a regular pattern (see Fig. 5–6). They appear during a particular hemisphere's late spring or summer, and most of them appear on the west sides of the oceans, apparently because these low-pressure cyclones are pushed westward by the trade winds and are affected by the Coriolis effect, thus swinging away from the tropics as they near land. You will note that there are no tropical cyclones in the South Atlantic or along the west coast of South America. This is probably because there is no opportunity for a storm of this nature to build up in the space that is available.

Review and Discussion

1. Explain the term *hydrologic cycle.*

2. How important is a region's water balance?

3. What is the difference between the processes of evaporation and sublimation?

4. Explain the meaning of *relative humidity.*

5. When can condensation occur?

6. How does fog form? Explain the difference between ground-inversion fog and advection-radiation fog.

7. Explain the difference between condensation and precipitation.

8. How are clouds classified? Explain the differences between cirrus, stratus, and cumulus clouds.

9. What does it mean when the word *alto* is attached to a cloud's classification, such as *altostratus?*

10. What does *nimbus* mean? What does *cumulonimbus* mean?

11. Explain these terms: *orographic, convectional,* and *frontal,* with reference to precipitation.

12. Explain the difference between cold front, warm front, occluded front.

13. What does *rain shadow* mean?

14. Explain why the southeastern portion of the United States receives more rainfall than the northeastern portion.

15. Explain the polar front theory.

16. Describe the formation of a thunderstorm. How does it differ from a tornado? How does it differ from a hurricane?

17. What parts of the United States are most affected by tornadoes? What parts are most affected by hurricanes?

18. Why is the South Atlantic Ocean spared the onslaughts of hurricanes?

Notes

[1] Morris Tepper (May 1958), "Tornadoes," *Scientific American* (offprint no. 848), p. 6.

[2] U.S. Department of Commerce (1970), *Tornado,* National Oceanic and Atmospheric Administration publication no. NOAA/PI 70007 (Washington, D.C.: U.S. Government Printing Office).

Additional Reading

Court, Arnold (August 1970), *Tornado Incidence Maps*. U.S. Department of Commerce ESSA Technical Memorandum ERLTM–NSSL 49. Norman, Okla.: National Severe Storms Laboratory.

———, and Richard D. Gerston (1966), "Fog Frequency in the United States," *Geographical Review,* 56, no. 4, 543–550.

National Oceanic and Atmospheric Administration (July 1972). *The Flood That Strikes in a Flash.* NOAA publication no. 511-323/164. Washington, D.C.: U.S. Government Printing Office.

Petterssen, Sverre (1969). *Introduction to Meteorology*. 3d ed. New York: McGraw-Hill.

U.S. Department of Commerce (1969). *Clouds*. ESSA brochure no. 0-355-259. Washington, D.C.: U.S. Government Printing Office.

6

THE EARTH'S WATERS

Water has three natural forms: liquid, solid, and vapor. With changes in form, it has recycled itself constantly since the creation of the earth.

From the time, hundreds of millions of years ago, when water vapor formed earth's first clouds and the first rain fell, this process of recycling has continued with little change in the quantity of the earth's water.

Every drop of the earth's original water supply is still in use . . . in the earth's atmosphere, on the earth's surface, or beneath it.

RIVER OF LIFE. WATER: THE ENVIRONMENTAL CHALLENGE
U.S. DEPARTMENT OF THE INTERIOR

It is probably obvious by now that all places on the earth are not fortunate enough to have a favorable water balance. The reason for this, as Table 6–1 shows, is the unequal distribution of the waters that make up earth's hydrosphere. Even a brief examination of this table will make it apparent that, although we cannot readily drink the salt waters of the oceans, they are the greatest suppliers of the moisture required to feed the hydrologic cycle and thus furnish the earth with the waters it must have in order to survive. Because our future may depend on how we treat them, we cannot ignore the waters of the earth. We must learn more about them so that we can devise methods to prevent abuse and pollution.

The Atmosphere

At times, when the skies are crowded with storm clouds and rain is falling, it almost seems as if the sky were filled with water. Then the storm passes, and the days that follow are clear and dry. At such times, we realize that, while the lower atmosphere does cover a considerable amount of territory—encompassing the earth as it does—it is sadly lacking in moisture. The troposphere contains a small amount of water at any given time, but it does its part in carrying out the hydrologic cycle. One might say that the lower atmosphere serves as a kind of middleman, as it takes water from the oceans, carries it inland, and then precipitates it.

Moisture in the atmosphere is held in the form of a gaseous vapor that is distributed throughout its tremendous volume. Hydrologists tell us that if all of this vapor were precipitated at one time and fell to the earth as rain it would form a layer only about 1 inch (2.5 centimeters) thick. How, then, can Mt. Waialeale in Kauai, Hawaii, average about 472 inches (1180 centimeters) of rainfall per year? The reason is the constant movement of air masses over the island. As one air mass is emptied of its moisture, a new one moves into the area, and the cycle continues. The atmosphere does not have the opportunity to become fully saturated at any one time: It is always giving and receiving. Consider the conterminous United States, which receives an average of 30 inches (75 centimeters) of precipitation every year and loses about 21 inches (53 centimeters) of this moisture by evapotranspiration.* As for the difference of 9 inches (23 centimeters), some of the moisture is obviously absorbed by the air over the land and reprecipitated.

Evapotranspiration is the total loss of moisture by evaporation from land and sea and transpiration from vegetation.

TABLE 6-1 World's estimated water supply

LOCATION	SURFACE AREA		WATER VOLUME		PERCENTAGE OF TOTAL WATER VOLUME
	Square Miles	Square Kilometers	Cubic Miles	Cubic Kilometers	
Surface Waters					
Freshwater lakes	330,000	858,000	30,000	123,000	.009
Saltwater lakes and inland seas	270,000	702,000	25,000	102,400	.008
Average in stream channels	—	—	300	1229	.0001
Subsurface Waters					
Soil moisture			16,000	66,000	.005
Groundwater within depth of 0.5 mile (0.9 kilometer)	50,000,000	130,000,000	1,000,000	4,096,000	.31
Groundwater—deep-lying			1,000,000	4,096,000	.31
Icecaps and glaciers	6,900,000	17,900,000	7,000,000	28,672,000	2.15
Atmosphere below an elevation of 7 miles (11 kilometers)	197,000,000	512,000,000	3100	12,700	.001
World oceans	139,500,000	363,000,000	317,000,000	1,298,432,000	97.2
Totals (rounded)			326,000,000	1,335,600,000	100.0

SOURCE: Adapted from U.S. Geological Survey Publication No. 2401-2189, 1972.

Soil Moisture

If you were to examine soils all over the world, you would find that almost all of them contain some moisture. "Some" moisture, however, is not enough for most forms of vegetation. If an area is deficient in moisture, artificial irrigation methods must be introduced for agriculture to be successful. What happens to the natural vegetation if moisture is lacking? Plants that have adapted to such an arid climate would survive—if the next arrival of moisture from precipitation or through groundwater movement is not too long delayed. Desert plants may flourish in an area lacking moisture, because of their ability to store water in their tissues, and because they transpire less than other vegetative forms. Transpiration can seriously affect the availability of soil moisture, and some trees may withdraw and transpire as much as 50 gallons (190 liters) of moisture per day.[1] For natural vegetation, this demands an energetic hydrologic cycle and frequent rains; for trees that are a part of mankind's agricultural system, repeated irrigation is required.

Silts and clays and other fine-textured soils with an adequate supply of humus (decaying vegetation) have the greatest potential for retaining moisture. They may take a long time to reach a level near saturation, but they also resist heavy losses. Sands and other coarse soils may reach saturation quickly, but they also lose their moisture rapidly through downward movement by gravity, as well as through evaporation from the surface. There is a great variety of soils, and each responds in a different way to a favorable or unfavorable water balance. In chapter 12 we shall investigate in depth the soils of the earth.

Groundwater

While it is likely that most people have some knowledge of the importance of underground water, it is also probable that they do not understand the true and vital significance of this source of water. While Table 6–1 makes it clear that most of the water we require for drinking and agriculture comes from lakes, rivers, or streams, we cannot discount the supply contained in the massive underground reservoirs. Compare the figures given for the water volume. All locations of surface water total a scanty 55,300 cubic miles (227,000 cubic kilometers), while the water volume of the subsurface water to a depth of a mere 0.5 mile (0.8 kilometer) totals 1,016,000 cubic miles (4,162,000 cubic kilometers). Consider also the percentage of surface water in the total water supply of the earth. The oceans, icecaps, and glaciers account for more than 99 percent of the total, and of the remaining amount (less than 1 percent), two-thirds is contained beneath the surface.

Groundwater has long been important to mankind. The ancient Babylonians worshipped a god called Ea, who ruled the domain of "springs that gushed from the earth,"[2] and the Judeo-Christian Bible tells how Moses "struck the rock with his staff twice, whereupon water in abundance gushed out, and the community and their cattle drank."[3] There is no doubt that fresh water issuing out of the ground

Mt. Waialeale, Kauai, Hawaii: The Wettest Mountain on Earth.

was believed to be the gift of God. People residing in arid or semiarid regions could not exist without wells, and history is full of tales of those who tried every means, including magical devices and divining rods, to find water. Only in relatively recent times have hydrologists understood the mysterious movements of groundwater.

THE MOVEMENT OF GROUNDWATER Moisture penetrates the groundwater area by infiltration from streams, ponds, lakes, melting snow and ice, and stream runoff. It moves by gravity. Water will infiltrate **porous** rocks (rocks having a great number of spaces between crystals or grains, which allow water molecules to enter), and it will move through rocks that are **permeable** (rocks in which water molecules not only are allowed to enter, but are allowed to move between the crystals and grains from one part to another).* If a layer of rock has both of these properties, and if it is also underlain by impermeable rocks that prevent the waters from infiltrating further downward, is termed an **aquifer.** An aquifer is much like an underground tunnel, because the waters may flow freely from high to lower elevations through the many spaces between the rock grains.

When water infiltrates the upper layers of the groundwater zones (see Fig. 6–1), it can move in several different directions. Downward flow is caused by the pull of gravity (called **gravitational movement**), and this is a continuing action, for as the bottommost layer becomes saturated, the water begins to flow laterally (more or less horizontally) by the force of gravity. Water can also move vertically upward by **capillary action,** which is caused when surface tension is reduced and the water molecules are drawn up through the tiny spaces within the rock or soil, or when plants reach down with their root systems to tap, or "suck up," the water. Not all of the water molecules will move in either direction, for some are held in place by what is termed **hygroscopic attraction.** This occurs when the water molecules coat or cover soil particles with a film and therefore cannot move.

The topmost zone, called the **zone of soil water,** supplies most of the short-rooted vegetation with moisture. This is also the area in which most of the plant nutrients can be found. Its moisture content fluctuates constantly, losing moisture through evaporation from the surface and via transpiration from its vegetative cover, and gaining moisture during times of precipitation.

As the amount of water increases, the molecules begin to gravitate downward into what might be termed the intermediate zone, but is more properly called the **vadose zone,** or the **zone of aeration.** This zone is aerated; that is, there is room for both water and air in sufficient quantities to support the larger forms of vegetation. Directly beneath this zone is the **water table,** which is the upper boundary of the lowest of the groundwater zones, the **zone of saturation.** The water table can be found close to or near the earth's surface (in humid areas or during seasons of

---

*The most porous rock layers are composed of sedimentary rocks, rocks built up from the deposits of sediments from lakes and oceans or from winds and running water. Those composed of fine particles, such as clay or shale, are porous enough to allow water to enter, but the spaces are too small to allow much vertical or lateral movement. Large-grain rock layers, such as sandstone, are both porous and permeable. Volcanic lava, if many cracks and fissures are present, can also be quite porous.

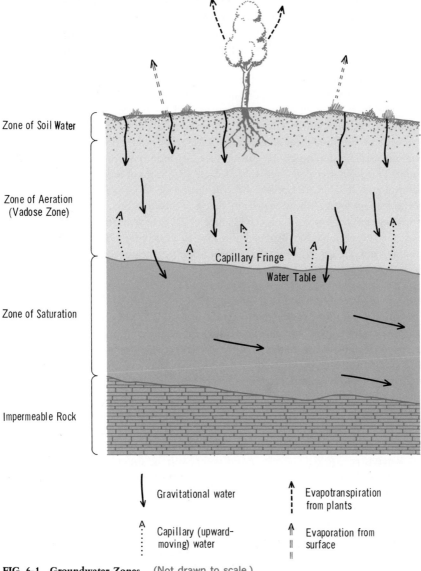

Zone of Soil Water

Zone of Aeration
(Vadose Zone)

Capillary Fringe

Water Table

Zone of Saturation

Impermeable Rock

↓ Gravitational water

A Capillary (upward-
moving) water

↑ Evapotranspiration
from plants

⇑ Evaporation from
surface

FIG. 6–1 Groundwater Zones. (Not drawn to scale.)

exceptionally high precipitation) or as deep as 100 yards (90 meters) during dry periods. The zone of saturation is the region where most of the underground water is to be found—some 1 million cubic miles (4,096,000 cubic kilometers)—as Table 6–1 indicates. At the bottom of this zone is, in many places, the earth's bedrock and, in others, dense rocks with only slight permeability. When water penetrates to this depth, it can only fill up the spaces between grains and crystals and, eventually, move down and away. The water table is sometimes referred to as the **interface,** or **contact surface,** between the region of soil moisture and the saturated layer, and it tends to parallel the ground surface above it.

The slope of the water table from its highest point of elevation, where it receives its initial charge of water, to its lowest point some distance (whether long or short) away, where the water is lost through discharge into lake, stream, ocean, or plain, is called the **hydraulic gradient** (Fig. 6–2). The amount of water flowing along this gradient is of the greatest importance, but much depends on the height of the source. The higher the elevation, the heavier will be the flow. If the water must travel a long distance, some will be lost en route, and the flow will be diminished. The height of the intake point will vary with the seasons, rising during times of maximum precipitation, lowering during times of minimum precipitation, times of drought. The vegetative cover will show telltale symptoms of lushness when moisture is plentiful and drooping, limpness, and wilt during the drier periods. As for streams in the area, Fig. 6–2(*a*) shows a stream that is supplied primarily by groundwater when the water table is at its highest. Figure 6–2(*b*) shows a streambed that is dry because the water table has dropped too low to replenish it. If there were water running in this stream, it would have to be supplied by surface runoff from some distant place where precipitation still occurs.

WELLS It has been estimated that close to 25 percent of the freshwater supply for our urban areas comes from underground sources. *Wells* are, simply, holes dug in the ground that are deep enough to go below the water table and into the saturated zone. In dry-land climate regions, you may have to dig very deep, and in some humid climate regions, you may only have to dig deep enough to form a basin in which the water can accumulate. The water table in China some 2600 years ago must have been very deep, for Confucius is reported to have said that if you dig a well to a depth of 72 feet (21.6 meters) and stop before reaching the spring, you have, after all, thrown away the well. In the present-day United States, the depths of wells are increasing, primarily because of our increasing populations. As more and more water is removed via wells, more wells must be dug, further depleting the water supply. Some places, such as in the city of Los Angeles, have had to reach out great distances to find the water needed to replace that taken from the ground. The problem had its beginnings nearly 70 years ago with the aqueduct that stretched some 180 miles (288 kilometers) from Owens Lake, situated at the foot of the Sierra Nevadas, to the city. With the increasing demand, water was brought to the Los Angeles area from Hoover Dam via a great canal through the Imperial Valley. Now, the Feather River Project has been created to bring water from the rivers near Sacramento. In addition, Los Angeles hydrologists take advantage of periods of heavy rainfall and divert much of the water from the rivers to catchment basins, where the water is temporarily impounded so that it will infiltrate into the aquifers and replenish, at least for a while, the groundwater reservoirs.

Another kind of a well, the **artesian well** (from the French province of Artois, where the first wells of this kind were dug) does not require pumps to withdraw the water, because the force of gravity and the buildup of hydrostatic pressure behind

(a)

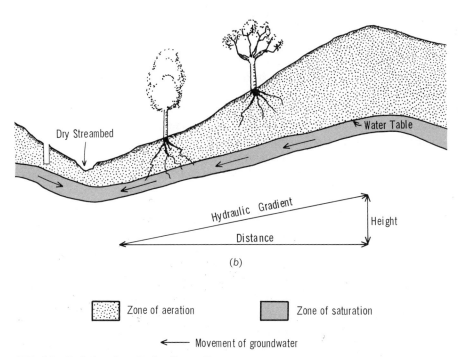

(b)

| | Zone of aeration | | Zone of saturation |

⟵ Movement of groundwater

FIG. 6–2 Variations in a Hydraulic Gradient.

the confined groundwater causes the water to erupt through the well opening in what sometimes seems to be a spouting fountain. As an example of the importance of artesian wells, consider the vast livestock region of Australia's Great Artesian Basin (Fig. 6–3). This basin is bordered on its east by the Eastern Highlands and on the west by the plateau that contains the Great Sandy Desert. An aquifer composed of sandstone dips downward from the highlands and stretches westward, beneath an overlying cap of impermeable shale, to the borders of the plateau, where it rises. Beneath the aquifer, which is over 1000 miles (1600 kilometers) in width, is impermeable bedrock. Water that falls on both plateau and highlands flows down through the aquifers and collects in the basin. When wells are dug, water escapes upward from its place of confinement.

SPRINGS AND GEYSERS   Springs may be either hot or cold. Cold springs result when an aquifer breaches the surface and water flows down a slope. Some springs flow continuously, others only intermittently, but they are always a welcome sight to a farmer or rancher—or to a backpacker hiking in the wilderness on a hot, dry day.

Hot springs and geysers may appear to have little commercial value, except as health spas or vacation resorts, such as at Yellowstone National Park in Wyoming and Hot Springs, Arkansas, but scientists have already found ways to harness the tremendous power of steam expanding from heated rocks beneath the earth's surface. This new method of supplying heat and energy for civilization's needs is called **geothermal energy,** and plants are currently in operation in Iceland, Italy, New Zealand, and the United States.[4]

Geothermal energy, as well as erupting geysers, are the result of water moving down through cracks and fissures in the earth's crust and coming in contact with superheated rocks near the upper layer of the mantle (see Fig. 9–3). The water is heated until it expands and turns into steam, then it rises toward the surface. The steam may bubble through standing bodies of water, erupt as a great, spouting fountain, sometimes at regular intervals (like Old Faithful in Yellowstone), or be harnessed by the equipment of a geothermal plant. Not all of these heated waters have a delightful aroma, especially if there happen to be large quantities of minerals present in the surrounding rocks. Although tourists at Yellowstone may hold their noses at the stench, commercial producers of chemicals and salts are happily planning to harvest these minerals, which will accumulate in the evaporation ponds at the geothermal plants.[5]

MAN-MADE AQUIFERS   For thousands of years, man-made aquifers have been around, but it is probable that most people think only of the concrete and steel water mains or irrigation canals in use today. Some may recall reading about the aqueducts built by the Romans several thousand years ago, but the most interesting man-made aquifers are those devised by Persians. These people left the broad and relatively lush steppes of Central Asia some 4500 years ago and migrated to the arid plateau of Persia. They were primarily cattle and horse breeders, but they practiced primitive agriculture as well, so they required water for their feed crops and for

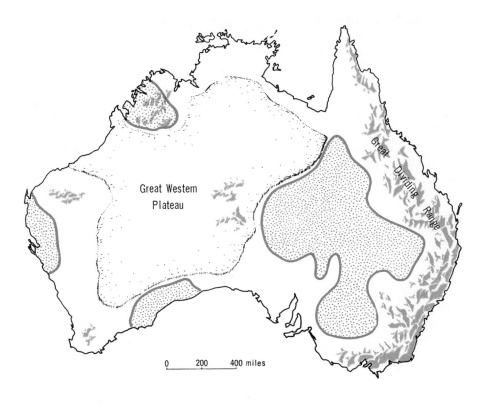

Great Western
Plateau

Great Dividing Range

0   200   400 miles

	Areas of artesian groundwater		Bedrock
	Edge of plateau		Surface materials
	Mountains		Artesian well
	Impermeable shale		Aquifer

FIG. 6–3 Australia's Great Artesian Basin.

Old Faithful in Yellowstone National Park: Geyser in Action.

themselves and their animals. This was a problem in an arid land where evaporation far exceeded precipitation. At first, they dug wells, but these too often filled with minerals and salts and were useless after a short time. They tried constructing irrigation ditches from the distant Tauros-Zagros Mountains, but the flowing water either seeped into the parched ground or rapidly evaporated into the dry air. Quite often villages died, and the inhabitants were forced to move on, hoping for new land where water would be plentiful. Then, some of these horsemen of so long ago came up with the answer: the **qanat.***

Essentially, a qanat (kanat) is a well that is nearly horizontal, rather than vertical. In order to get water for a distant village, a team of water-finding experts located a water-bearing aquifer. Then a team of "engineers" and "diggers" dug a well next

*To learn more of this interesting method of water transport, see: George B. Cressy (January 1958), "Qanats, Karez, and Foggaras," *Geographical Review* 48, 27–44; also, Anthony Smith (1953), *Blind White Fish in Persia* (New York: Dutton).

to the mouth of the aquifer, usually located in the upper levels of a broad alluvial fan at the mouth of a valley. This first well was dug to a depth of about 100 feet (30 meters) and was called the mother well. A line was laid out toward the distant village, and the digging began in earnest. Shafts, or wells, were dug at regular intervals, usually about 50 yards (45 meters) apart, to serve as ventilation tubes and to allow the excavated soil and rocks to be removed. At the top of each shaft a mound was formed, giving the land above the qanat a characteristic appearance—like a series of doughnuts laying on the land. Then began the arduous job of connecting the shafts by means of an underground tunnel or conduit (Fig. 6–4). The tunnel, roughly rectangular in shape, averaged from 2 to 3 feet (1 meter) in width and was up to 6 feet (1.8 meters) in height.

When the qanat was completed, the aquifer was breached, and the water began to flow down through the length of the tunnel to the village. Some of these qanats in use today measure up to 50 miles (80 kilometers) in length and still supply the villages with water. During the rainy season in the western mountains, some qanats may deliver as much as 4200 gallons (16,000 liters) per minute, diminishing to only 400 gallons (1500 liters) per minute during the dry season.[6] One qanat is generally adequate for 200 acres (80 hectares), enough to support the average Iranian village.[7] It is estimated that there are between 20,000 and 50,000 qanats still in use today.[8]

FIG. 6–4 Cross Section of an Iranian Qanat.

 Aquifer   Impermeable rock

Streams, Rivers, and Lakes

We have seen in Table 6-1 that the earth's groundwater reservoirs contain about 36 times as much water as the surface locations, but those figures do not tell yhe entire story. It may be obvious that groundwater is vital to mankind, serving as a supplier of fresh water for plants, animals, and humans, but surface waters are important, too. Consider the relationship of humans to a deep lake that is well stocked with fish and a full-flowing river that provides easy access to the hinterland, as well as to the sea. Consider also the tremendous variety of agricultural, commercial, and industrial activities that depend on surface water for the generation of power. Think of the recreational activities made possible by the construction of giant hydroelectric dams and the resulting man-made lakes that sometimes stretch for miles behind them.

Another significant consideration is the role played by surface waters in the hydrologic cycle. It has been estimated that the 66 principal rivers of the world discharge about 3720 cubic miles (15,237 cubic kilometers) of water every year.

The Transforming Power of a Giant Hydroelectric Dam.   The landscape of the Upper Danube in the Federal Republic of Germany has been changed because of this modern installation. Besides supplying water and power, the dam is a crucial link in the creation of a navigable waterway from Germany to the Black Sea.

The estimated total from all rivers, large and small, measured and unmeasured, is about 9200 cubic miles (38,000 cubic kilometers) yearly, which is 25 cubic miles (104 cubic kilometers) daily. At any given moment, however, only about 300 cubic miles (1228 cubic kilometers) is physically present in all the world's stream channels, and it seems evident that river channels, on the average, contain only enough water to maintain their flow for about two weeks.[9] It also seems that groundwater comes to the aid of streams, supplying them with water during times of minimum precipitation.

Figure 6–5 shows some of the most important rivers of the world, rivers that made their mark not only on human history, but on the face of the earth as well. River flow is governed by the amount of precipitation or additions from melting glaciers or snowpacks at the river's source; consequently a regular flow is never certain—with the possible exception of Egypt's "Holy Mother" Nile. This river, swollen by tropical rains and meltwater from the Ruwenzori Mountains and the Ethiopian Highlands, rises by late spring and overflows its banks, flooding both sides for considerable distances. By fall, the waters subside, leaving a thick deposit of fertile silt, a "gift of the goddess" to the farmers. So regular was the flooding of the Nile that the ancient Egyptians devised their 365-day year from the Nile's clocklike precision.[10]

The Nile, however, does not match rivers such as America's Mississippi, which has a drainage area of some 1,243,000 square miles (1,988,800 square kilometers) and discharges at an average rate of 620,000 cubic feet (18,600 cubic meters) per second. This amounts to some 133 cubic miles (545 cubic kilometers) per year, which is almost 34 percent of the total river discharge from the United States. Compare these figures to those of the mighty Amazon River of Brazil, which discharges almost 1300 cubic miles (5324 cubic kilometers) per year—almost 3 times that of all the rivers of the United States.[11]

It becomes readily apparent that the tremendous discharge of water in the form of **surface runoff** (movement of water across the land) to the oceans is of considerable importance in maintaining the hydrologic cycle. In addition, the waters from the rivers add to the groundwater reservoir and supply moisture via evaporation to the atmosphere.

Lakes, which number in the hundreds of thousands, are also important in maintaining the hydrologic cycle. While every lake, large or small, is valuable to someone, the large freshwater lake merits our closest attention. Figure 6–5 brings out an interesting fact: The large freshwater lakes are to be found only in the northern one-third of North America, in east-central Africa, and in Siberia. This suggests that in the other areas of the world, the water is either in rivers or underground. The estimated total volume of all the large freshwater lakes comes to about 30,000 cubic miles (123,000 cubic kilometers), with a combined surface area of some 330,000 square miles (858,000 square kilometers). Some of these lakes are impressive in area as well as in volume. Others, such as Siberia's Lake Baikal, may not have the area of the others, but do surpass them in volume (Fig. 6–6). Baikal resulted from a mighty fracture of the earth's crust, which opened a deep cleft in the ground (a process we shall consider in more detail in chapter 10). The depth of Lake Baikal

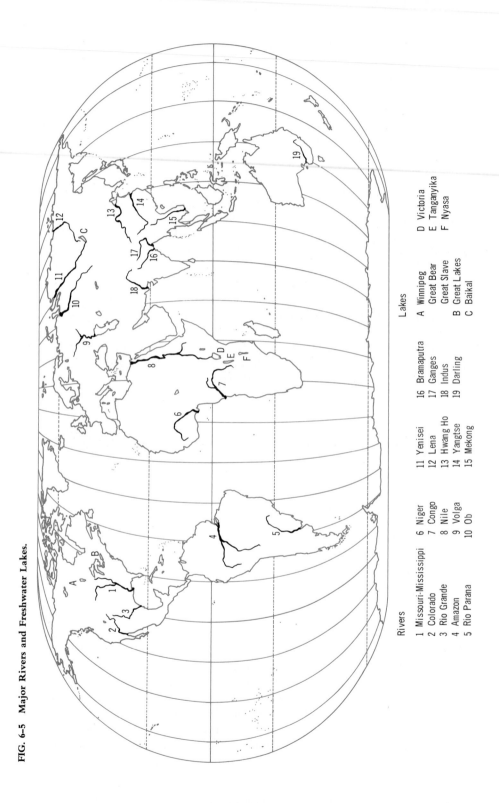

FIG. 6-5 Major Rivers and Freshwater Lakes.

Rivers

1 Missouri-Mississippi
2 Colorado
3 Rio Grande
4 Amazon
5 Rio Parana

6 Niger
7 Congo
8 Nile
9 Volga
10 Ob

11 Yenisei
12 Lena
13 Hwang Ho
14 Yangtse
15 Mekong

16 Bramaputra
17 Ganges
18 Indus
19 Darling

Lakes

A Winnipeg
B Great Lakes
C Baikal

Great Bear
Great Slave

D Victoria
E Tanganyika
F Nyasa

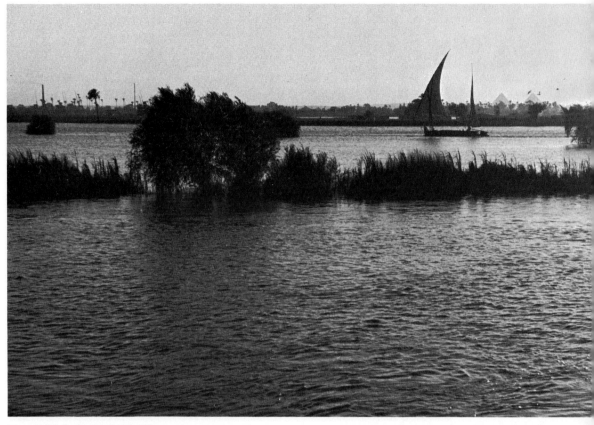

Egypt's "Holy Mother" Nile River.

has been measured to be 4500 feet (1350 meters), making it the deepest lake in the world. However, the five Great Lakes that lie on the border between the United States and Canada, if considered as a body, constitute the largest area of fresh water in the world.

One final note on the freshwater lakes: They account for about 75 percent of all the fresh water on the earth. North America's amount is 26 percent; Africa's, 29 percent; and Asia's, 21 percent.

As to the saltwater lakes and their position in the scheme of things, in addition to supplying fish and other products of the sea, their primary service to mankind is as waterways. The Caspian Sea is the largest of all, having a surface area of 152,084 square miles (395,418 square kilometers) and a water volume of 19,240 cubic miles (78,800 cubic kilometers) and accounting for over 75 percent of the saltwater lakes. Most of the saltwater (saline) lakes are to be found in Asia. The only one of any significance in North America is Utah's Great Salt Lake, which has shrunk to only 1700 square miles (4420 square kilometers) and a water volume of 7 cubic miles (29 cubic kilometers).

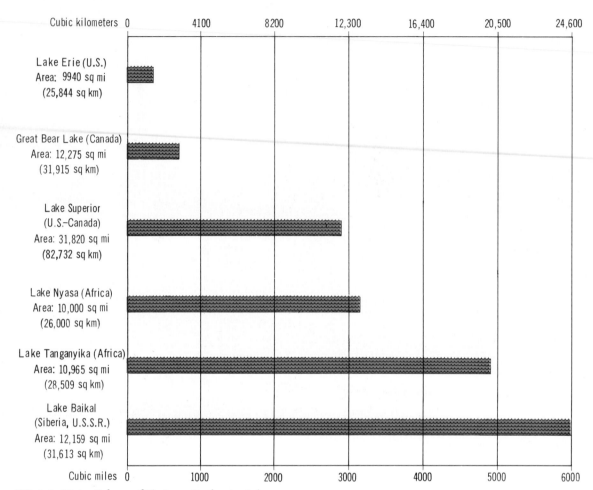

FIG. 6-6 Water Volume of Six Large Freshwater Lakes.

Icecaps and Glaciers

Slightly over 2 percent of earth's moisture is held in glaciers and **icecaps** (Fig. 6–7). This may seem insignificant when contrasted with the 97 percent held by the world's oceans, but when contrasted with the amounts held by streams, rivers, lakes, and groundwater (all of which total a mere 0.6 of the world's water supply), it is a significant amount.

Glacial ice comes in many forms, from relatively small glaciers tucked away in remote valleys in high mountains, such as the Alps, the Canadian Rockies, the Andes, the Pamirs, the Tien Shan, and the Himalayas, to huge masses of ice, such as the icecap that presses down on the island of Greenland, the floating ice sheet of the Arctic, and the vast **continental glacier** that covers Antarctica. The smaller **valley glaciers** (also called **alpine,** or **mountain, glaciers**) serve the purpose of sculpturing the

▲ Alpine (valley) glaciers

▨ Icecaps, floating ice pack, continental glaciers

FIG. 6-7 Earth's Icecaps and Glaciers.

mountainous regions of the world and supplying the inhabitants of the farmlands below their terminals with water, as well as having a profound influence on local weather conditions.

However, as a major control of weather and climate for immense regions, we must look to the great masses of ice in the polar regions, which, today, have retreated to relatively limited areas. In the past, however, the earth has experienced great ice ages—long periods when ice sheets spread down from the North Pole onto the continents of North America and Eurasia, grinding away at the earth, scouring out huge basins that were later filled with water (giving us the Great Lakes, for example), laying bare the crust's bedrock, sending out powerful rivers that deposited fresh residues of soil and silt over tremendous areas, and lowering the level of the oceans in the process. All of this occurred during times of high temperature and high evaporation that caused the troposphere to become so covered with clouds that incoming solar radiation was turned aside, thus allowing the temperatures below to drop to very low levels. When precipitation occurred, rather than falling as rain, it fell as snow and ice. The earth has gone through the cycle of ice age and interglacial age again and again, with the last of the great ice ages having ended some 6000 to 10,000 years ago. No one can foretell if and when another will happen.

The world's valley glaciers hold a water volume of about 50,000 cubic miles (205,000 cubic kilometers), which is less than that held by lakes and rivers, far less than that held by the Greenland icecap—630,000 cubic miles (2,580,000 cubic kilometers), and a mere fraction of that held by Antarctica—6,320,000 cubic miles (25,800,000 cubic kilometers). What would happen if all the valley glaciers were to melt? If this occurred, the localities below them would experience flooding of their lands and some alterations in their weather and climate. If, however, the

Antarctica: The Largest Remaining Continental Glacier.

Antarctic continental glacier were to melt, the level of the oceans would rise, perhaps as much as 100 feet (30 meters) or more, possibly inundating the lower elevations of such cities as New York, Galveston, London, and Leningrad. Fortunately, such an occurrence is unlikely, at least in our time, but the remote possibility does point up the importance of the immense reservoir of water retained by the world's expanses of ice.

The Oceans

One almost has to agree with Leonard Engel, who said that those ancients who gave our planet its name, Earth, were really all wet.[12] Their concern was with the continental landmasses, with little importance placed on the oceans. They thought the earth was all land, with ocean merely meandering about the outer edge. Today, however, we know that the oceans make up almost three-fourths of the earth's surface and account for over 97 percent of the water supply. Obviously, they play a major role in the workings of the hydrologic cycle and exert a tremendous influence on the climates of the world.

DISTRIBUTION OF THE OCEANS Examining a globe, you might at first conclude that there is just one big ocean, with a scattering of landmasses here and there. Technically, you would be absolutely correct. There is only one global ocean, but it has been arbitrarily subdivided into four primary oceans. We know these as the Arctic, Pacific, Atlantic, and Indian oceans (Fig. 6–8), but where does each start and stop? The nations of the world, working through the International Hydrographic Bureau (headquarters in Monaco), have devised limits that arbitrarily separate the oceans from each other; these separations, however, are "primarily for the purpose of filing *Notices to Mariners* and have little to do with natural boundaries."[13] Table 6–2 shows the comparative areas covered by the oceans, and shows that the Pacific Ocean covers more than twice the area of any other ocean and almost as much as all the other water bodies (including the multitude of seas, bays, and gulfs that surround the continents) combined.

Because each ocean does have its own currents, as well as varying amounts of warm and cold waters, it is possible to go further and subdivide the four oceans into seven bodies of water: North Pacific, South Pacific, North Atlantic, South Atlantic, Indian, Arctic, and Antarctic. As we proceed with our discussion of the oceans, the relevance of such classifications will become apparent—with the possible exception of the Antarctic Ocean.

Most oceanographers do not accept the Antarctic as a separate ocean. Rather they consider it merely the southern extension of the Pacific, Atlantic, and Indian oceans. However, oceanographic research conducted during the International Geophysical Year (IGY) has convinced many that there is sound reasoning for classifying it separately. There is a distinct body of water that is in constant movement about the continent of Antarctica, a powerful cold-water current that circulates from west

FIG. 6-8 Oceans of the Earth.

Arbitrary dividing lines established by the International Hydrographic Bureau. 20°E longitude separates Indian and Atlantic Oceans; 147°E longitude separates Indian Ocean from the Pacific; 80°W separates the Pacific from the Atlantic.

Line of separation between warm waters of the major oceans and the cold current of the Antarctic

TABLE 6-2 Comparative sizes of the world's oceans

OCEAN	AREA IN SQUARE MILES	AREA IN SQUARE KILOMETERS
Pacific	64,000,000	166,000,000
Atlantic	31,700,000	83,000,000
Indian	28,400,000	74,000,000
Arctic	5,400,000	14,000,000
Subtotal (rounded)	129,500,000	337,000,000
Sea, bays, and gulfs not included in above	10,000,000	26,000,000
Totals (rounded)	139,500,000	363,000,000

to east between latitude 40°south and 60°south. This circumpolar current is encouraged (indeed, it is dragged along) by the prevailing westerly winds of the region, as well as by the absence of any obstructing landmasses. When we discuss the circulation of the ocean currents, you will see the importance of this whirling mass of cold water and recognize why all the oceans exert such an influence on the climates of the world.* First, however, we must answer the question of *why* the waters of the oceans move from one part to another, much like great rivers. The answer can be found by examining the composition of the ocean waters and by keeping in mind the effect of the world's wind systems.

THE NATURE OF OCEAN WATERS The waters of the world ocean behave in a manner similar to the air masses: They move horizontally and vertically, and they move because of differences in density. **Density** is defined as the weight of a specific volume of water at a specific temperature, expressed in grams per cubic centimeter. For example, water reaches its highest density at a temperature of 39°F (3.9°C). At this point, the weight of 1 cubic centimeter of water would equal 1 gram on a balance scale. The density of any ocean water body is determined by its temperature and its salinity.

Salinity In effect, **salinity** means "saltiness." If you have ever gone swimming in an ocean, you know that, in comparison to the water of a river or lake, the ocean seemed salty. In actuality, ocean water is almost 96.5 percent "pure." That is, only 3.5 percent of the total is made up of the 57 different kinds of minerals (such as traces of gold, arsenic, nickel, silver) and salts (such as sodium, chlorine, sulfate, magnesium, calcium, and potassium). When sodium and chlorine combine, the result is sodium chloride, more commonly known as table salt. To harvest this important salt, ocean water must be impounded in a collecting basin, and after the water has evaporated, the salt left behind can be gathered and refined. What would happen

---

*For an interesting report on the oceanographic research conducted in the Antarctic during the IGY (1957–58), see: V. G. Kort (September 1962), "The Antarctic Ocean," *Scientific American.*

if all the waters of the oceans were to evaporate? How much salt would we find? The answer shows you just how big the oceans are: The layer of salt would cover the entire earth to a depth of about 160 feet (48 meters)!

Salinity of the oceans varies from place to place, depending on such factors as the amount of precipitation, rates of evaporation, temperatures, and the movements of the waters. Figure 6–9 shows that high salinities can be found in various places along the subtropical high-pressure belt. The Atlantic, especially the region north of the Tropic of Cancer, rates high in salinity, while most of the Pacific, particularly the deeper layers, is low in salts. The lowest salinities are found in the polar regions, owing to the constant addition of melting snow and ice and a very low evaporation rate. The highest salinities are found in water bodies that essentially are cut off from the general circulation of the oceans, such as the Red Sea, the Persian Gulf, and the Mediterranean. This is because higher salinities will be found wherever the rate of evaporation exceeds precipitation, or wherever fresh or less salty water is prevented from entering the body and thereby diluting it. Generally, in the midlatitude regions lower salinities will be found, because although precipitation is down, so is the rate of evaporation. In the tropics, the rate of evaporation is high, yet salinities are low. This is because there is considerable precipitation in these areas. As for the Persian Gulf and the Red Sea, they have high salinities, yet they are situated relatively close to the tropics. Here, there is a high evaporation rate, and the narrow inlets to these bodies of water do not allow dilution by waters from the Indian Ocean. Further, those areas are located in the midst of the world's greatest desert region (the Sahara, Arabian, and Iranian deserts), where there is little precipitation and are no freshwater rivers.

Contrast that situation with the complex one to be found in the Gulf of Mexico, located at about the same latitude. The Gulf of Mexico averages about 36‰ (3.6 percent) salinity, but near the mouth of the Mississippi River the salinity drops to about 20‰ (2.0 percent).* The reason for this is that the mighty river not only brings salts to the Gulf of Mexico but also dilutes the saltiness by dispersing the salts throughout the area and adding fresh water to the total supply. Waters that are high in salinity are heavier than less salty waters, and they tend to sink toward the ocean bottom, where they then begin to flow out and away from the source. A good example of this can be seen in the interesting, if paradoxical, movements of the waters of the Mediterranean Sea. The waters are high in salinity, owing to the high rate of evaporation. Thus they tend to sink down to the bottom and flow westward through the only outlet available: the Straits of Gibraltar. At the same time, the less salty, thus less dense, waters of the eastern Atlantic flow eastward into the Mediterranean. The Atlantic waters flow near the surface, passing over the dense waters of the Mediterranean, which flow close to the bottom.

People once believed that the oceans received their salts from the ocean floor, yet we now know that the salts and minerals come from the land, carried by

---

*Salinity is expressed as a ratio, in terms of the "parts per thousand" of the dissolved salts as related to a specific volume of water. For example, the average salinity of sea water is 35 parts per thousand (35‰), which means that there are 35 grams of dissolved salts for every 1000 grams of water. A high salinity would be 40‰; a low salinity would be 33‰.

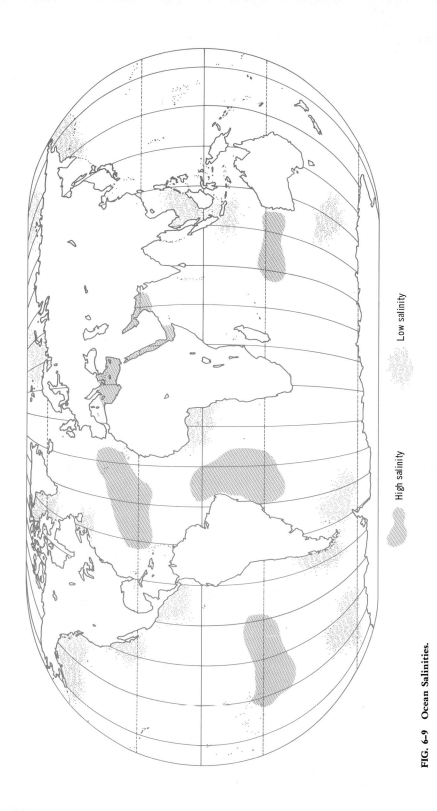

High salinity

Low salinity

FIG. 6–9 Ocean Salinities.

groundwater and rivers. This may seem like a strange paradox, but the salts arrive within a new supply of relatively fresh water and, although the salts do accumulate through time, they are spread both vertically and horizontally throughout the entire body of water.

Temperature As in the atmosphere, temperature plays an important role in the oceans. The waters on the surface are heated by insolation and radiation from the atmosphere to a depth of about 6 feet (1.8 meters). Winds and currents cause a mixing of this surface water with waters below, resulting in a layer of relatively warm water that measures from 600 feet (180 meters) to 1500 feet (450 meters) in thickness. This layer is often referred to as the **mixed layer,** a band of constant temperature. Below this mixed layer is a region called the **thermocline,** which means that there is a rapid decrease in temperature, extending downward to a depth of about 3500 to 6000 feet (1050 to 1800 meters). Below the thermocline, the water temperature decreases more slowly through the deep waters to the ocean floor.

Where are the warmest waters? Look again at Figs. 4–4 and 4–5, isothermal maps of the world in January and July, and you will see that the receipt of insolation is heaviest on and near the equator and the lower latitudes and lowest in the upper midlatitudes and the polar regions. Compare these maps with Fig. 6–10, a generalization of the world's ocean currents, and you will see that the regions of high insolation send warm water flowing poleward, while cool currents stem from the regions of low insolation and high albedo. Generally, warm currents flow on or near the surface, while the cool currents slide along or near the ocean floor. This occurs because the water molecules in the layer of relatively high temperatures separate in their excited movements; thus, the water becomes lighter, less dense. In cool or cold waters, the molecules slow down and become tightly compressed, making the waters denser and heavier. Dense water tends to sink and flow outward from the source, and warm water tends to rise in a convective movement similar to heated air. Then the winds, aided by the Coriolis effect, take over and steer the currents in the patterns you saw in Fig. 6–10.

Currents Oceanographers often compare the circulation of the oceans to that of the earth's wind systems, for just as ascending or descending air masses cause lateral movement of the air masses they displace, so do moving bodies, or **currents,** of water. As an example, you will note that warm currents sweep westward under the equatorial belt of low pressure, urged along by the trade winds, then part, sending one warm current northward, one southward toward the poles. At the same time, cool waters from the polar regions flow toward the east then swing toward the equator. Warm waters move along the western sides of the oceans, cool waters along the eastern sides. As an example, consider the circumpolar Antarctic current. This cold-water current has an average surface temperature of around 28°F (-2.2°C), because insolation is low and the albedo is high in the south polar region. The cold waters sweep around the continent, and "rivers" branch off the main stream and slide northward beneath the warmer waters further north. The cold waters eventually rise, or upwell, near latitude 30° south, along the coasts of three continents. This

80°N

60°N

40°N

20°N

Equator

20°S

Gyre

40°S

60°S

Gyre

◀—— Cool currents   ◀---- Warm currents

FIG. 6-10 Generalized System of Ocean Currents.

creates deserts such as the Atacama of Chile and Peru, the Namib of Southwest Africa, and the Australian Desert. Cool currents originating in the Arctic also have a hand in the formation of deserts in the Northern Hemisphere.

The waters travel great distances in their circumnavigation of an ocean, and a state of balance is reached, with cold water moving out of the polar region and warm water entering somewhere further along to replace it. There is constant interaction of warm and cold waters, and these moving masses of water act much like moving weather fronts. Just as low-pressure cells race out of the polar fronts to bring chilly weather and storms to the midlatitudes, so the cool ocean currents influence the weather and climate of places like Greenland, Iceland, Alaska, Siberia, and, somewhat surprisingly, the Sahara Desert of North Africa. As tropical cyclones, given birth by the warm and moist air masses of the equatorial regions, drive north to enter the Caribbean and the southeastern portion of the United States, so warm ocean currents flow along the eastern sides of the continents, bringing warmth and moisture to what otherwise might be parched and cold lands.

An interesting sidelight to the pattern of current flow in the Atlantic Ocean is the existence of the **gyres** (sometimes called **gyrals**) that are situated under subtropical high-pressure cells. These gyres result from the inward-spiraling of currents that spin off warm, poleward-moving currents (such as the Gulf Stream) and circle around, dragged by the winds of the subtropical high. While not all the proof is in, the waters of these gyres are believed to "bulge" up to a level 4 feet (1.2 meters) higher than the waters surrounding them.

The North Atlantic gyre is famous in myth and legend as the "graveyard of lost ships" and is a swirling region of thick masses of seaweed. Portuguese sailors named it the Sargasso Sea. An intriguing and mysterious portion of this gyre is a somewhat triangular section of the West Atlantic, which stretches west from Bermuda to the east coast of the United States, then south to the Virgin Islands. Since 1866, some 12 ships and 60 aircraft have disappeared, leaving no traces—no debris, no oil slicks, no survivors, and no explanations.[14] Some have theorized that they may have encountered sudden storms at sea, powerful turbulence in the sky, magnetic interference, attacks by German U-boats (during wartime), or "unidentified flying objects." John Godwin, in *This Baffling World,* reports that there was never any radio communication that implied the presence of danger, and the trouble always "seemed to stem *from direction and location.*"[15]

MOVEMENTS OF OCEAN WATERS   While differences in temperature and salinity have an important influence on the movement of ocean waters, it is the wind that is most responsible for the major ocean currents. We are not referring to local winds, but to the major and prevailing wind systems. The relatively steady blowing of the trade winds, both north and south of the equator, starts the waters moving from east to west, but—like the winds themselves—the ocean currents are deflected by the Coriolis effect, and currents in the Northern Hemisphere tend to veer (curve) to their right (Fig. 6–10), while those in the Southern Hemisphere veer to their left. A circling pattern is set up—aided by other factors, such as a continent that gets in the way—forcing the waters to pile up and to seek avenues of escape. Opposing currents, those that run in the opposite direction of the prevailing system, may be observed flowing beneath the surface.

The idea that such equatorial **countercurrents** exist has long been recognized, but it has only been in recent times that oceanographic research has given us accurate details. For example, the Pacific's equatorial countercurrent was studied by Townsend Cromwell of the United States Fish and Wildlife Service in 1952. The Cromwell current, as it is now termed, flows eastward from New Guinea, along the equator, to the coast of Ecuador. Its depth reaches from 100 to 750 feet (30 to 225 meters), and it moves at speeds up to 3 knots.[16] The Atlantic also has a countercurrent that flows back out of the Caribbean and along the northeastern coast of South America toward Africa. It is believed that there also is one in the Indian Ocean, but this has not yet been proved.

Major ocean currents The major ocean currents are shown in Fig. 6–11, and the pattern we saw earlier in Fig. 6–10 can be seen repeated again and again, from ocean to ocean: a clockwise spiraling in the water bodies of the Northern Hemisphere, counterclockwise in the Southern Hemisphere. When we discuss the earth's vegetation, you will see more clearly the tremendous impact of the warm and cool ocean currents. As an example, the eastern portions of Asia and North America are under constant onslaught by low-pressure cells, emerging from the polar front, that could leave the land cold and dry. Fortunately, however, Asia has the Japan current (also known by the Japanese name *kuroshio,* "black current"), which is born in the north equatorial current of the Pacific. This warm current sweeps along the Asian coastline, modifying the climate as it goes, supplying plentiful moisture and warm temperatures to eastern China, Korea, and the islands of Japan. A similar situation exists in the North Atlantic, with the north equatorial current giving birth to the Gulf Stream, which warms the eastern United States and, as the North Atlantic drift, brings a pleasant climate to the British Isles and Scandinavia.

The portion of the Indian Ocean south of the equator matches the patterns of the other oceans, but the region north is noted for its complexities. The currents in the Arabian Sea and the Bay of Bengal circle in a clockwise motion during the summer, but they reverse their directions in the winter. This rhythm generally follows the prevailing wind system of the region; thus, the currents are called monsoon currents.

Wind waves The most dramatic and exciting of the ocean movements are the **wind waves** (with the possible exception of the seismic sea waves resulting from earthquakes occurring in the ocean floor, which we shall discuss in chapter 14). Anyone who has been to the seashore knows that the waters of the ocean are in constant, restless motion. The most obvious movements are the waves that come rolling in toward the beach, to end up as foaming breakers. These waves are rightly called *wind waves,* because they are caused by the passage of wind over the water's surface. The resulting transmission of wind energy to the water disturbs the once-placid waters, stirs them into dynamic action, and starts them rolling.

There are different kinds of wind waves, depending on a number of factors: depth of the water body, closeness to a shoreline, and the velocity of the wind and its duration. If the starting point, or source region, is in the open sea where the waters are very deep, wave movement follows a definite and rather regular pattern (Fig. 6–12). If you were on a ship in the open sea, the waves would appear to be "marching" in a particular direction, wave after wave. In actuality, however, the water is moving up and down in a circular movement. What you see as a moving wave is not the water that is moving, but the wave form and energy. To prove this point, take a 20-foot length of rope and tie one end to a post. Now, back off until the rope is fairly straight, but not too tight. Snap the rope, as if you were cracking a whip, and a "wave" will travel all the way through the rope to the post.

FIG. 6–11 Major Ocean Currents of the World.

1 North equatorial
2 South equatorial
3 Countercurrent
4 Gulf Stream
5 Sargasso Sea (gyre)
6 North Atlantic drift
7 Labrador
8 Greenland
9 Canaries
10 Brazil
11 Gyre
12 Falkland
13 Agulhas
14 Mozambique
15 Japan current (kuroshio)
16 North Pacific drift
17 Kamchatka
18 Alaska
19 California
20 Monsoon drifts
21 West wind drift (antarctic)
22 Humboldt (Peru)
23 Benguela
24 West Australian

Cool currents

Countercurrents

Warm currents

Direction of Wave Progression

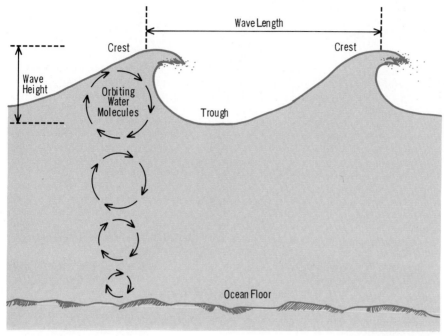

FIG. 6-12 Wave Motion. The curving line of the sea shows the rhythmic rise and fall of a wave as it is driven across the ocean by the blowing wind. Water molecules move in a circular path as they orbit about their focal point: up, forward to the crest, down, and back to their starting point. The water mass does not itself move any appreciable distance forward; only the wave energy and form moves. An object on the surface (a leaf, a branch, or a patch of oil) is physically transported ahead, pushed by the wind, but the water molecules remain behind, circling about their focal points. The diameter of the orbital path decreases with increasing depth and ceases entirely at the ocean floor. Actual forward movement occurs only in shallow waters, when the wave pattern is disrupted and the wave crest surges forward, carrying the mass of water molecules onto the beach.

Did the rope itself move? No, it only *seemed* to move; what you saw was the wave energy you transmitted to the rope.

Wind waves begin because of a disturbance (storm) at sea. Once the wave process is set in motion, either by the actual pushing by the wind or by friction (sometimes referred to as skin drag), the waves will proceed as long as either of the two forces continues to supply energy to the movement. The orbiting movement of the water particles gives added impetus to the speed of the wave, because as the circling energy rises to the crest and momentarily surges forward before it makes its descent, the wind's push and pull is increased. As a result, you may sometimes see waves that appear to be moving much faster than the wind. **Whitecaps,** tiny breakers at the top of the waves, appear if the wind blows at a faster rate than a moving wave.

Wind waves and Whitecaps.

When the wind waves have moved a great distance from their place of origin and from the winds that gave them birth, swells result, which may move in a direction different from that of the local winds. In calm regions, the marching waves lose speed and height and begin to "die." No longer will you see sharply pitching waves studded with whitecaps; instead, you see gently rolling waves that are rounded and smooth.

The height and length of a wave depend primarily on the initial speed and continuing velocity of the wind, how long it lasts, and the amount of ocean expanse to be covered (this distance is called the fetch). If the wind is strong and steady and the fetch is long, the waves may build to great size. There are verified reports of waves attaining heights of 90 feet (27 meters) and, on rare occasions, heights of 112 feet (34 meters).[17] During World War II, two destroyers were overturned and sunk by high waves in a storm east of the Philippines. Mariners have a rule of thumb that says the wave height will not go above double the wind velocity. If this is true, the winds would have to reach a velocity of at least 50 knots to create waves 100 feet (30 meters) high. Wind waves in the tropics are relatively low and short in length, except during storms that produce winds of gale force.

One of the most interesting of all the earth's regions, with reference to wind waves, is the area between 40° and 50° south latitude, an area aptly termed the "roaring forties." With no landmasses to interfere, the westerly winds attain speeds up to 80 knots! Imagine the sizes of the waves produced, and think of what Magellan, Sir Francis Drake, Captain Cook, and all the other intrepid mariners of the days of sailing ships experienced when they were forced to "round the Horn" from the Atlantic to the Pacific.

Tides Gigantic waves that overturn ships and drown coastal cities are not the result of tides. Rather, **tides** are the rise and fall of the sea level, owing to the influence of the moon's gravitational pull against the waters of the earth. The moon may be 240,000 miles (384,000 kilometers) distant, but it does have a vital influence on conditions on earth. The moon orbits around the earth (Fig. 6–13) and exerts a pull that is difficult to perceive in reference to the landmasses, but its presence is felt as far as the oceans are concerned. If the moon did not revolve around the earth, or if the earth did not rotate, we would have no tides at all. Fortunately, both of these do happen, and we have four tidal changes every 24-hour period: two high tides, two low tides, alternating every 6 hours (approximately) from high to low to high to low. The highest **tidal ranges** (the difference in height between high tide and low tide) occur when the earth, moon, and sun are in line with each other (called *syzygy*). These are called the **spring tides.** The lowest tidal ranges occur when the moon is "out of line" (at right angles to the line between the sun and the earth); thus the pull of the sun tends to cancel out the pull of the moon. These are called **neap tides.**

While the changing of the tides may not seem to affect our weather or climate, they do have an important role to play in many people's lives. When the earth is at perihelion (its closest approach to the sun), the pull of the sun on the waters of the earth is intensified. If this should occur during a spring tide, the resulting tidal range would be dramatically increased. If, also, a storm occurred at sea, sending powerful wind waves in to shore, the resulting surge could be catastrophic for people living along the shoreline. This did happen in January 1974, along California's lower coastline, when the extra-high tides were combined with storm-tossed waves to cause considerable damage to the houses along the shore. On the beneficial side is the prospect of constructing hydroelectric power plants in narrow river mouths to produce energy by harnessing the tidal flows.

There is much more to be said about the influence of tides, and we shall pursue this further in chapter 14, when we study the role of the oceans in sculpturing earth's landmasses.

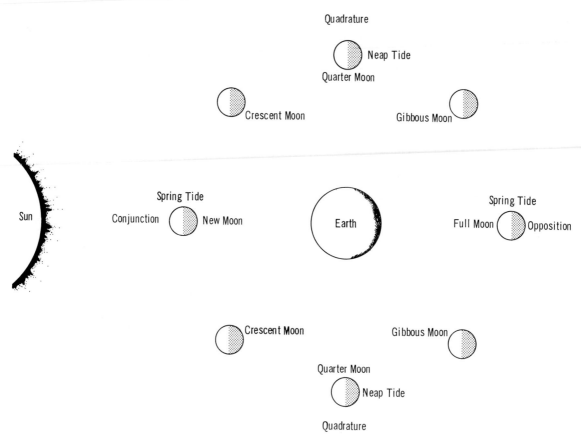

FIG. 6-13 Positions of the Moon and Their Effect on the Tides.

Earth's Waters: A Second Look

The waters of the earth are a global concern. At least, they should be, for the hydrologic cycle is not confined to any one area. It is a worldwide phenomenon that ties all of the lands and their peoples together. Consider the long journey of a drop of water that rises from the Pacific Ocean as vapor, precipitates over the Rocky Mountains, flows into the Gulf of Mexico, is transported by the Gulf Stream to Europe, where it continues the cycle: evaporating and precipitating, carried by groundwater movement, running streams, and blowing winds, until, at last, it rejoins the Pacific. And the cycle begins anew.

What one nation does to the waters, through carelessness or callousness, affects all of mankind. The human population has increased to such an extent that the hydrologic cycle has been seriously affected. Think of the growing number of large cities, massive urban areas covered with houses, buildings, streets, and freeways, and consider what this has done to the once-massive groundwater reservoirs, the atmosphere above the cities, and the streams, lakes, and seas nearby.

It is, perhaps, time for us to act, not just react, to work a little harder to save one of our most precious resources, water.

Review and Discussion

1. How does the earth's atmosphere compare to surface and underground water as a supplier of fresh water?

2. How important is soil moisture to mankind?

3. What is the difference between porosity and permeability?

4. What is an aquifer?

5. Explain the difference between gravitational water and capillary water.

6. Make a sketch of the groundwater zones, and explain the difference between the zone of soil water and the zone of aeration.

7. Explain the term *hydraulic gradient.*

8. What is an artesian well?

9. Explain the use of the Persian qanat.

10. What is meant by the term *surface runoff?*

11. Which is the deepest lake in the world? Which is the largest?

12. Explain the difference between these terms: *valley glacier, icecap, continental glacier.*

13. Which of the oceans is the largest? The saltiest? The coldest?

14. What does salinity have to do with the density of ocean water?

15. Explain the term *thermocline.*

16. Explain the term *gyre.*

17. What does *upwelling* mean?

18. What is meant by the term *countercurrent?*

19. How are ocean currents influenced by the Coriolis effect?

20. What makes the *kuroshio* (current) important to the Japanese?

21. What causes ocean waves?

22. Explain these terms: *fetch, whitecap, swells.*

23. What causes tides? What is meant by *tidal range?* Explain the difference between spring tides and neap tides.

Notes

[1] U.S. Geological Survey (1972), *Waters of the World* (leaflet no. 2401-2189), (Washington, D.C.: U.S. Government Printing Office), p. 14.

[2] *Larousse Encyclopedia of Mythology* (1959), trans. Richard Aldington and Delano Ames (London: Paul Hamlyn), p. 53.

[3] Numbers 20:11.

[4] Garry D. McKenzie and Russell O. Utgard, eds. (1972), U.S. Geological Survey, "Geothermal Energy," *Man and His Physical Environment: Readings in Environmental Geology,* (Minneapolis, Minn.: Burgess), pp. 232–234.

[5] McKenzie and Utgard, eds., Donald E. White, "Resources of Geothermal Energy and Their Utilization," pp. 235–239.

[6] George B. Cressy (January 1958), "Qanats, Karez, and Foggaras," *Geographical Review,* 58, 29.

[7] Cressy, p. 37.

[8] Cressy, p. 39.

[9] U.S. Geological Survey, p. 8.

[10] Luna B. Leopold and Kenneth S. Davis (1970), *Water* (New York: Time-Life Books), p. 121.

[11] U.S. Geological Survey, p. 6.

[12] Leonard Engel (1962), *The Sea* (New York: Time-Life Books), p. 9.

[13] Harold W. Dubach and R. W. Taber (1968), *Questions About the Oceans,* National Oceanographic Data Center publication G-13 (Washington, D.C.: U.S. Naval Oceanographic Office), p. 10.

[14] John Godwin (1968), *This Baffling World* (New York: Hart), pp. 235–254.

[15] Godwin.

[16] Dubach and Taber, p. 24.

[17] Willard Bascom (August 1969), "Ocean Waves," *Scientific American* (offprint no. 828), p. 5.

Additional Reading

Chorley, Richard J., ed. (1969). *Introduction to Physical Hydrology*. New York: Barnes & Noble.
——— (1969). *Introduction to Fluvial Processes*. New York: Barnes & Noble.
——— (1969). *Introduction to Geographical Hydrology*. New York: Barnes & Noble.
Cowan, Robert C. (1969). *Frontiers of the Sea: The Story of Oceanographic Exploration*. Garden City N.Y.: Doubleday.
Cromie, W. J. (1962). *Exploring the Secrets of the Sea*. Englewood Cliffs, N.J.: Prentice-Hall.
Darwin, Charles (1959). *The Voyage of the Beagle*. Abridged and ed. Millicent E. Selsam. New York: Harper & Brothers.
Ericson, David, and G. Wollin (1967). *Ever Changing Sea*. New York: Alfred Knopf.
Highsmith, R. M. (July 1965). "Irrigated Lands of the World," *Geographical Review,* 55, 382–389.
Horsfield, Brenda, and Peter B. Stone (1972). *The Great Ocean Business*. New York: Coward-McCann & Geoghegan.
Ley, Willy (1962). *The Poles*. New York: Time-Life Books.

Olson, Ralph E. (1970). *A Geography of Water*. Dubuque, Iowa: Wm. C. Brown.

Sayre, A. N. (November 1950). "Groundwater," *Scientific American* (offprint no. 3818).

Smith, Anthony (1953). *Blind White Fish in Persia*. New York: E. P. Dutton.

Thomas, H. E. (1951). *The Conservation of Ground Water*. New York: McGraw-Hill.

Todd, D. K. (1959). *Ground Water Hydrology*. New York: John Wiley.

Tolman, C. (1951). *Ground Water*. New York: McGraw-Hill.

Weyl, Peter K. (1970). *Oceanography: An Introduction to the Marine Environment*. New York: John Wiley.

7

EARTH'S VEGETATION

Climate exercises a very strict and often direct control on the development and distribution of plants, animals and soils. . . . On the other hand, climate itself is profoundly affected by the nature of the vegetation cover; wind speed, temperature regime and atmospheric humidity just above the surface of the ground beneath the shade of dense forest are remarkably different from those at just the same height above soil level in nearby grassland.

VEGETATION AND SOILS: A WORLD PICTURE, S. R. EYRE

So far, we have considered a number of important cycles: the cycle of the changing seasons as the earth revolves in its orbit about the sun; the cyclic heat balance reached between the sun and the earth's atmosphere; and the hydrologic cycle. These cycles are important to the continued operation of the systems that affect our environment: the atmosphere (air) and the hydrosphere (water). We shall now examine the third great system, the biosphere.

The **biosphere** (from the Greek *bio,* meaning "life") is sometimes referred to as the life zone, for it is the realm within which earth's life forms are able to exist. Its base begins somewhere within the upper reaches of the earth's crust, and it extends an unknown distance out through the lower atmosphere. There is a close relationship within this zone among all the elements—soils, vegetation, moisture, air temperature, and the varied life forms—and earth scientists refer to the biosphere as a **biotic complex**, or **ecosystem**.[1] This means that a change affecting one element of a system can affect all the others. Our immediate concern is with one element of the ecosystem: vegetation and its cycle of life.

The Vegetative Cycle

The **cycle of vegetation** is, to many people, a wondrous phenomenon. In early spring you may have paused to wonder at the new buds bursting forth from the once-naked limbs of a tree in your yard. When this happened, you may have discovered within you a growing feeling of awe, a feeling of reverence—a feeling that you were witnessing the creation of life. Then, after a long, hot summer, comes the fall season. Commercial crops are harvested, and the natural vegetation around you begins to show signs of weariness. Grasses and shrubs wilt, and leaves turn color and fall from the trees. You may then have experienced a feeling of sadness, as though you were watching death reach its hand over the landscape. Did you then realize that you had observed the beginning and the end of a life cycle? Well, if you experienced these feelings, you were not alone, because such sentiments have been an integral part of the life experiences of many individuals throughout the history of mankind.

RELIGION AND THE VEGETATIVE CYCLE

The religions evolved by the peoples of ancient times speak plainly of their reverential attitude toward the grasses, shrubs, and trees they knew and of their belief that each person was a part of nature, not above it. As an example, think of the cattlemen and horsemen of the steppes of

Central Asia who roamed those wide pastures some 4000 or 5000 years ago. They called this land "the sea of grass," and it meant life to them and their herds. If the sky father did not send sufficient rain, or if the sun god did not provide enough days of warmth, the grasses died and, then, so did the cattle and horses. Many times, when conditions of drought lasted for long periods, the tribes were forced into long marches, searching for greener pastures. They soon came to associate the bountiful produce issuing from the earth as "gifts" from the earth mother. Perhaps they did not truly understand the relationship between sky and earth, nor the reasons behind the changing of the seasons, the death and rebirth of the sun, or the difference between day and night. To them, it was all a mystery related to the capriciousness of gods and goddesses, a capriciousness that must, somehow, be placated. The Slavic ancestors of the Russians believed that a diminutive god named Leshy ruled the forests. The Babylonians worshipped a sun god called Shamash and hoped that he would take care of them and the world about them. The peoples of the hot and humid rainforest regions of Africa, South America, and the East Indies were more concerned with the spirits of the rivers and the trees than with a supernatural figure who ruled the skies. Peoples of middle and high latitudes were forced to endure harsh winters, and it seemed to them that nearly everything died during that period when the icy north winds screamed down upon their tents or villages. There was, however, one exception. The sturdy evergreen trees withstood everything the storms threw at them and so came to be a symbol of eternal life.

An ancient and beautiful legend tells the story of Balder, son of the great god Odin. Balder the Beautiful, the best-beloved of the Germans and Scandinavians, was the intermediary between the sky gods and mankind, but he was slain by the evil Loki, Prince of Darkness, who had discovered that the only living thing that had not pledged fealty to Balder was the humble parasite, mistletoe. A fire-hardened dart of this normally limp and fragile plant killed Balder, sending him into the underworld. Odin's plea to the god of Hel that mankind needed Balder was only partially granted: Balder could return to the earth's surface at the end of winter, but he must return to Hel each fall.*

Similar stories throughout the religions of mankind point to the idea that the rebirth of life each spring is associated with the rebirth of a god or, at least, the result of the will of a god. Primitive agriculturists also recognized the changing of the seasons, for life was always more comfortable and easier for them during the time of spring, summer, and early fall. Some plants seemed to have more persistence than others, and these came to have religious significance, such as the aforementioned evergreen of the Germans, the lotus of the Indian Buddhists, and the maize that was reverenced by the Aztecs and Maya of Meso-America.

The earth's natural vegetation has changed considerably since those early times, changes made by human hands. It is to be hoped that we can regain some of the old concern in the years to come.

---

*One of the most interesting discourses on the close interrelationship between mankind and vegetation is Sir James George Frazer's (1960) *The Golden Bough* (New York: Macmillan).

Religion and the Vegetative Cycle. (*a*) is a rendering of Leshy, forest divinity of the ancient Slavs of Russia. (*b*) shows Shamash, Babylonian judge of the heavens and the earth, god of the sun, worshipping the sacred tree of life. In (*c*), Balder, revered by Scandinavians and Germans as the messenger from God, lies dying, struck down by an arrow made from mistletoe. Among those gathered around Balder are warriors of Valhalla, the great king Odin, and evil Loki, the Prince of Darkness.

(*a*)

(*b*)

(*c*)

SUCCESSION AND CLIMAX VEGETATION   The type of vegetation that abounds in any given area is determined by the physical environment, which, in turn, is the result of a number of factors: the parent material from which the soil is derived, the amount of precipitation, the topography of the land, the insolation and air temperatures, and the latitude. If no drastic changes have been wrought by man, the hooves and teeth of animals, or vagaries in the climate, a particular plant community will evolve. This vegetation will be "just right" for the region. It will be the **climax,** the culmination of a series of plant communities that strove to adapt to the environment. The climax vegetation will have established a balance, a state of equilibrium, with the environment.

The term **succession** refers to the changes in the plant communities, from the first types that appeared in the region to the climax. As an example, let us consider one of the Hawaiian islands, the so-called big island of Hawaii. The island is volcanic in origin, so all of the soils have the same beginning—but not the same ending. The northeastern slopes are on the windward side and exposed to heavy precipitation; the southwestern slopes are deficient in precipitation but receive higher amounts of insolation. Suppose seeds are dropped by high-flying birds, an equal amount on each side. Will the two plant communities evolve in the same way? Not likely, for if the seeds are from plants that require considerable moisture but do not need great amounts of sunlight, those dropped on the windward slopes will thrive, those on the leeward slopes will die out. As time passes, more seeds are either washed ashore or deposited by birds. Perhaps this time the seeds are from plants that need high temperatures and can exist on little moisture. The seeds deposited on the southwestern slopes will take root; the other seeds will not be able to adjust and will disappear.

If, however, we consider an area that is bare of all other forms of vegetation, new types will try to adapt. They will take root, struggle, die out. Then, in succession, other types will try to become established. Some will succeed, and some will not. Eventually, only those that can survive in the climate will remain. Thus, the climax is reached.* The competition among plants for living space is quite severe, and some plants that might have survived are crowded out or "strangled" by invading plants that are better able to make a rapid adaptation to the climate and the soils. A good example of this can be seen in the Tuolumne Meadows area of Yosemite National Park. Forest fires and the intrusion of civilization reduced certain sections of coniferous forests (pines and spruces) to charred rubble. Grasses immediately appeared and filled the vacant areas. In time, however, the conifers rebounded and reasserted their "right" to the meadows. Now, the meadows are receding, and the forest is advancing on all sides.

What happens when humans enter the picture? When the first Viking set foot on the soil of New England and gazed about him at the thick forests, was he seeing the original climax vegetation of that region? What of the vegetation that confronted the ancestors of the Australian aborigines when they first arrived in the land "down under"? Archaeologists tell us that the Sahara of northern Africa may seem barren today, but it was not always this way. Until 1500 B.C. it was covered with great

*Some plant geographers use the term *climatic climax vegetation,* thus emphasizing the influence of climate in the establishment of the true natural vegetation for any given area.

expanses of grasses and forested areas. Perhaps in this instance, mankind should not be held responsible, for other factors may have been more instrumental in changing the vegetation that existed in those regions.

ENVIRONMENTAL CONTROLS   The main controlling factors of environment are: moisture, air temperature, topography, soils, and biological activity—including mankind. All are integral parts of the biosphere, and it is difficult to classify them in order of importance.

195

Moisture   A potent factor in the development of plants is moisture. While plants must adapt over time to varying conditions, the presence or absence of water is a strong determinant. Some plants, such as seaweed, water lilies, and sedges (called **hydrophytes,** from *hydro,* "water," and the Greek *phyton,* "plant"), can live under water and can survive at depths of about 15 feet (4 to 5 meters) because of their ability to utilize sunlight at those depths and thus allow the process called **photosynthesis** to proceed.* Other plants can live in seemingly impossible regions, such as deserts, where the precipitation is almost zero. If moisture-loving plants are transplanted to those regions, they will quickly die, even though the deserts do have occasional rainstorms in which rain falls in a hurried downpour and quickly soaks into the ground or rushes off through *wadis* (normally dry streambeds similar to gullies). Sometimes there are seeds just waiting for this opportunity. They have been lying underground, dormant and apparently lifeless. With the addition of this new moisture, they quickly germinate, and the desert blooms for a short period. Seeds are produced that fall into the soil and wait their turn for next year's brief rainshowers.

Deserts lying within reach of warm, moist maritime air masses gain a little more precipitation, and this allows a considerably different type of vegetation to evolve. The Sonoran Desert of northern Mexico and the American Southwest may seem rather barren at first glance, but a closer look discloses an exciting variety of plants that can thrive in a hot desert that has only 5 to 10 inches (12 to 25 centimeters) of precipitation: the giant saguaro cactus, the Joshua tree, plus many other species of cacti.

Around the edges of the dry deserts are grasslands called prairies and savannas. Although the air temperatures in these regions are generally high, precipitation of over 10 inches (25 centimeters) is able to support vast expanses of varying types of grasses and, primarily in the savanna regions, occasional clumps of trees or shrubs.

Forested regions require plentiful precipitation, and some of them receive over 45 inches (112 centimeters) of rainfall annually. Where the forests are composed of towering trees with heavy foliage, sunlight will not be able to reach ground level, and undergrowth will be scanty or nonexistent.

Air temperature   In conjunction with the availability of moisture, air temperature operates to determine the size of plants, their leaf structure, their root systems, and

* *Photosynthesis* is the process used by plants to manufacture the food they need. They take carbon dioxide out of the air—or the seawater—and, with the aid of sunlight, use the chlorophyll within their tissues to transform the carbon dioxide into the carbohydrates (sugar and starch) they require.

(a)　　　　　　　　　　　　　　　　　　　　　　*(b)*

(c)　　　　　　　　　　　　　　　　　　　　　　*(d)*

The Effect of Water on the Evolution of Plant Communities.　　*(a)* hot, dry desert: average annual rainfall less than 5 inches (12.5 centimeters). Cover: occasional plants after rainshowers. *(b)* hot, semiarid desert: average annual rainfall from 5 to 10 inches (12.5 to 25 centimeters). Cover: succulents such as cactus, Joshua trees, small and scattered shrubs. *(c)* hot, wet and dry grasslands of the savanna type: average annual rainfall over 10 inches (25 centimeters). Cover: grasses, occasional shrubs and trees. *(d)* hot and moist forests; average annual rainfall over 45 inches (115 centimeters). Cover: permanent stands of trees with little undergrowth.

whether they can grow in clumps, in large stands, or singly. Temperature is important in controlling the growth of all plants, because it affects the chemical activity (movement of nutrient-bearing sap and the process of photosynthesis), periods of dormancy, production of seeds, and the amount of time needed for an adequate growing season. If temperatures are high, evaporation from the soils will be high, as will transpiration from the plant itself. If moisture is lacking, plants in this type of area will be forced to adapt by developing leaves that are thick and spongy in order to retain the moisture. If both temperature and precipitation are high, plants will cover the earth surface, and the leaves may be thin, perhaps filmy, and transpiration rates will be high.

Temperature can also limit the spread of certain types of plants. The July 50°F (10°C) isotherm (see Fig. 4–5), for example, is the limit of tree growth in the Northern Hemisphere. Beyond this line is the tundra region, a land of lichens and sedges that can survive in soil that is permanently frozen (**permafrost**) beneath the surface of the ground. The same control is asserted in mountainous regions where the **timber line** reflects the temperature boundary above which trees cannot grow.

Topography   The shape of the land, its **topography,** is also important as a control in the evolution of plant communities, especially with regard to the **slope** and the **aspect** of the area. While there are many areas of the earth that are fairly flat and level, most have some slope. If the terrain is steep and rugged, soils have a difficult time evolving because of their inability to retain water, which tends to run off swiftly. If slopes are exposed to constant winds, there will be an increase in evaporation and thus less moisture retention. Flat bottom lands may be unable to rid themselves of excess moisture, and swamps or bogs may result.

Aspect refers to the direction a slope is facing in relation to the sun. Slopes facing the sun's rays will receive more insolation than slopes facing in the opposite direction. As an example, consider the accompanying photograph, which shows a section of mountainous terrain in the Northern Hemisphere. The area facing the south has a light cover of grasses and clumps of shrubs; the area facing north is covered with heavy vegetation. If both areas receive about the same amount of precipitation, why does one slope have heavier ground cover than the other? For one thing, moisture is more easily retained on the north-facing slope because there is less insolation and thus a lower rate of evaporation.

Near the equator or in the lower latitudes, slope is more significant than aspect, because all slopes will receive about the same insolation. If precipitation is heavy, soils may be washed down from higher elevations, thus limiting the growth of plants along ridges and creating steeper slopes. In tropical areas with high temperatures as well as high precipitation, the cover of vegetation may be rather dense on all mountain slopes. If a timber line exists, it will be at a much higher elevation than on mountains located in the mid- or high-latitude areas. Low-lying areas in hot, moist climatic regions will be swampy or marshy if there is not enough slope to carry away the excess moisture. In such an area, the water table is usually high and close to the surface, so the soils tend to be waterlogged.

Topography as a Factor in the Development of Plant Communities. Notice the difference in aspect: The north-facing slope has heavy vegetation; the south-facing slope has very light vegetation.

Soils Another effective determinant in the development of plant communities is the soil (a subject that will be covered in detail in chapter 12). If a soil has had time to evolve from its beginnings as a part of the earth's bedrock in a humid region, and if it contains sufficient amounts of **humus** (decaying plant material) or nutritious chemicals, it will be able to retain more moisture and, thus, to support a luxuriant vegetative cover. On the other hand, soils evolving in arid or semiarid areas (where evaporation exceeds precipitation) may have a considerable amount of humus, but they are loose-grained and therefore more porous, thus allowing moisture to drain away rapidly. Many plants in dry regions are drought-resistant and are referred to as **xerophytic** (from the Greek *xeros,* "dry").

Biological activity The term **biological activity** refers to the action of living organisms, such as bacteria, worms, rodents (such as gophers and moles), and quite definitely, humans. Humus, mentioned earlier as being necessary for a good vegetative cover, results from the activity of bacteria, which cause decomposition of leaves, blades of grass, limbs (and even the entire trunks) of trees, as well as bodies of dead animals. When precipitation occurs, the water soaking into the ground carries the humus particles with it, thus enriching the soil. Worms and small burrowing animals move about through the uppermost layers of the soil, manufacturing air spaces and making channels for water to infiltrate by gravity to deep-lying roots of shrubs and trees. All of these factors contribute to the evolution of soils, but the same cannot always be said for the activities of mankind. With the ever-increasing population and the subsequent spread of urbanization, streets and highways are paved, houses and buildings are erected, the native water supply is depleted, and the soils are sometimes excessively robbed of their nutrients. As an example, cattle ranchers moved into the Great Plains of the United States and stripped the land of its climax vegetation and broke up the *sod cover* (the upper soil layer that is tightly knit because of the extensive root systems of the native grasses). These "sodbusters" of the period from 1870 to 1930 practiced serious overgrazing of the land, which resulted in the "dust bowl" of Kansas and Oklahoma.[2]

199

The Structure of Plants

Natural vegetation is "wild" vegetation, not yet tamed or altered by human intervention. It is to be differentiated from domesticated varieties (which we might term cultural vegetation), and it varies from region to region all around the world. We have discussed the controlling factors of the environment, and it would appear that climate is the most important of all. The climate of one place is not necessarily the climate of another, for other factors must be considered, such as the position of an area relative to mountains, deserts, plains, rivers, lakes or oceans, and winds. All of these controlling factors combine to create plant forms that differ from each other in the structure, shape, and texture of the leaves and their coverage, size, and periodicity.*

LIFE FORMS OF PLANTS   There is great variation in the life forms of plants, from trees like the towering *Sequoia gigantea* (more commonly known as the giant redwood) down to the minute, simple-celled systems called algae and fungi. Trees are the largest of all plant systems and are not difficult to distinguish from other plants, as they have a single trunk from which, usually, many branches are put forth. Shrubs are next, tough, woody plants that are hardy, low-growing systems. They are not limited to one trunk but have several stems sprouting from the nodule and root base. Another type of woody plant is the liana, a plant system that has many climbing vines. Included in this group are wisteria, cucumbers, peas, poison ivy, and beans. Some of these climbers pull themselves up with stems that operate like tiny fingers, others use thorns as grappling hooks, and others wind their tendrils around the plant they are ascending and so move higher and higher.

The next group of plants, grasses, are uncomplicated systems with simple leaf structures and tubular stems. Some grasses, including corn, wheat, rye, oats, and barley, which might be called food grasses for humans, and some, such as fescue and Kentucky bluegrass, are used for animal food and lawn covers. Whether annual or perennial, grasses die back to the ground after flowering and producing seeds. Related to the grasses are herbs, plants that also lack woody stems and may be either annual or perennial in their life cycles.

*For a detailed description of plant classification by structure, see Pierre Dansereau (1957), *Biogeography: An Ecological Perspective* (New York: Ronald Press).

(a)

(b)

(c)

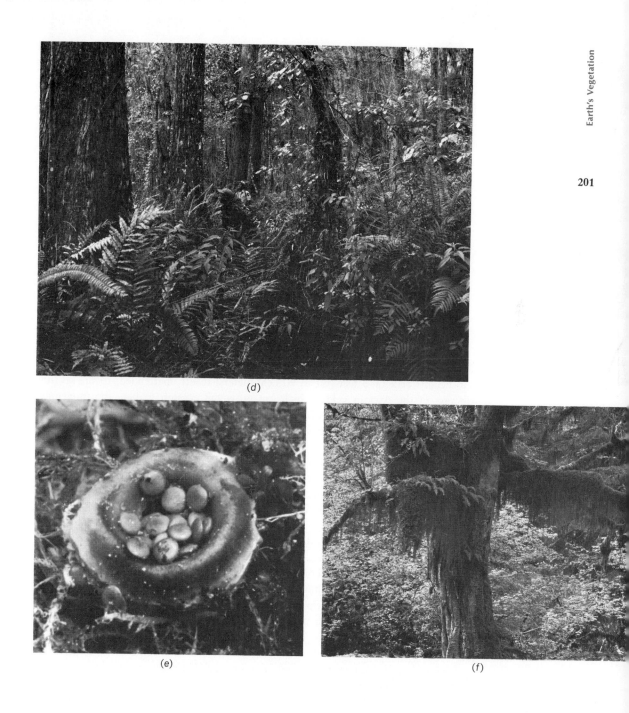

Vegetation Life Forms. (*a*) algae formed on a body of still water; (*b*) giant sequoias (redwoods); (*c*) grasslands in Denmark; (*d*) ferns and lianas in Corkscrew Swamp, Florida Everglades; (*e*) Bird's-nest fungus in the Hoh rainforest; (*f*) moss-covered maple on the Hoh River Trail (Olympic rainforest, Washington).

Plants that grow and thrive on top of other plant forms are called **epiphytes** (from the Greek *epi,* "upon"). Wind-carried seeds become lodged on branches or in the bark of trees and, in time, the seeds germinate. Epiphytes move through their cycle completely out of touch with the ground, yet they are not parasites—that is, they do not take nourishment from their host tree. The most common epiphytes are orchids, ferns, mosses, liverworts, and lichens, and they are usually found in the humid tropics.

At the very bottom of the plant scale are the fungi and algae, members of the simplest of all plant forms and systems, the **thallophytes** (from the Greek *thallos,* "young shoot or frond"). Many of the thousands of life forms within this system are single celled and, thus, have no need for a vascular network (like tiny blood vessels) to carry nutrients. Because algae live in water, they are able to absorb the required moisture directly through the walls of the cell. The most common forms of algae are diatoms and seaweed. Fungi are represented by, for example, bacteria, yeasts, mildew, and mushrooms and are considered parasitic—that is, they live on and draw nourishment from dead or living plants and animals. Neither algae nor fungi have root systems or stems. When algae and fungi live together, almost as one plant, the result is called a lichen.

Plant size is another way to identify different vegetation life forms, for plants range in size from the tiny, almost invisible bacteria to gigantic trees. For example, grasses vary greatly in size, from the kinds we use as lawn covers, which can grow to heights of 2 or 3 feet (60 to 90 centimeters) if left unattended, to the savanna grasses of the African veldt, which can grow tall enough to hide the movement of a herd of elephants. Consider also the grass called bamboo, which can attain heights of 100 feet (30 meters).

COVERAGE AND PERIODICITY Coverage can also be used to classify plants. Some trees grow singly, such as East Africa's acacia, while others grow in clumps surrounded by grasses. Still others flourish in great stands, as, for example, the teak forests of Burma, where the trees grow very close together, crowding out other plant forms. Climate sets the boundaries for the different systems, but competition for sunlight and space is also an effective determinant.

The term *life cycle* (also called *periodicity*) is used to express a plant's response to the cycle of the seasons. For example, some lawn grasses never seem to die, but continue growing (and needing to be mowed every week) throughout the year, while others grow rapidly through spring and summer, then die back when fall arrives. This also applies to shrubs, herbs, and trees: Some go dormant and lose their leaves; others seem to retain their leaves from season to season. The period of dormancy can be likened to a period of sleep when the life processes slow down, owing to lack of sunlight or precipitation, or when the temperature drops too low. Trees that drop their leaves during this time are called **deciduous.** Trees that appear to keep their leaves are called **evergreen.** The reason we say they only *appear* to retain their leaves is that, while the tree always has leaves, there is a continual shedding of old leaves to make way for the new. Some forests may appear to be evergreen, but actually have trees that take turns in dropping leaves throughout the year. Such forests are classified as *semideciduous*.

LEAF TYPES   Leaves of different plants vary considerably, providing another way to classify plant systems. There are two basic leaf types: **broad leaf** and **needle leaf** (Fig. 7–1). Needle leaves are found on conifers (cone-bearing trees) and certain species of cactus. In the case of the conifer, the needle is so constructed that a single vein runs through its center. This vein collects moisture and carbon dioxide from the air and also transports plant sugars to other sections. Needle leaves also resist excessive losses of moisture during periods of high temperatures and evaporation and are not subject to freezing. Broad leaves are found on flowering plants and come in many different sizes, shapes, and textures. The most common type of broad leaf is termed *membranous,* because it is a thin, flat layer of tissue of average thickness and is soft in texture. In contrast to the needle leaf, the broad leaf has many veins—main, secondary, and small netted veins—and all perform the necessary task of manufacturing food for the plant. Oxygen and moisture (by transpiration) are emitted by the leaf, and it has been estimated that 1 acre (0.4 hectare) of corn will transpire more than 300,000 gallons (1.1 million liters) of water during a single season.[3]

Other types of broadleaf plants, such as ferns, have thin, almost transparent leaves, which are called *filmy.* Plants such as holly and mesquite exist in a dry habitat, and the lack of plentiful moisture causes their leaves to be tough and thick. These are termed **sclerophyllus** (from the Greek *scleros,* "hard"). The thick, spongy leaves of a cactus are called **succulents** (from the Latin *succulentus,* "full of juice"), because they hold a considerable amount of water. These types evolve as a response to an area of light-to-scanty rainfall and a high rate of evaporation. To protect themselves from moisture loss, the plants evolve tough, reflective skins and spongy interiors.

Distribution of Natural Vegetation

Because there are no hard-and-fast rules governing the classification of the earth's natural vegetation, we shall establish only three major divisions and then formulate subdivisions as needed. Our classification will include the forest, grassland, and desert vegetation.

FORESTS   Figure 7–2 is a map of the earth's major forest regions: tropical rainforests, tropical deciduous, Mediterranean woodland and scrub, broadleaf summer deciduous, midlatitude coniferous, and the taiga.

Tropical rainforests, sometimes referred to as equatorial rainforests, are the most prolific, luxuriant, and extensive of all the forests on and near the equator. These forests are commonly referred to as *selva* (the Brazilian term for this type of woodland) and are typical through much of middle America (from Yucatan southward); in Venezuela, Colombia, Ecuador, and Brazil of South America; in portions of West

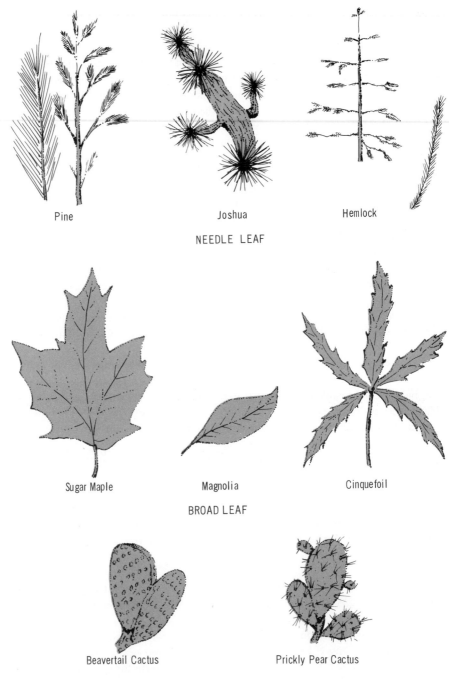

Pine   Joshua   Hemlock

NEEDLE LEAF

Sugar Maple   Magnolia   Cinquefoil

BROAD LEAF

Beavertail Cactus   Prickly Pear Cactus

SUCCULENTS

FIG. 7-1 Leaf Shapes and Textures.

Africa and Zaire; on the east coast of Malagasy; on the west coast of southern India; in Bangladesh and Southeast Asia; in the East Indies to New Guinea; and on the east coast of Australia.

Most of the trees are of the broadleaf variety. They appear to have evolved a workable system of stratification—that is, they "layer" themselves, and each level is allowed so much room and just so much sunlight. Therefore, adapting to its situation, a tree will grow to a particular height and no more. Because of the intense overcrowding, there is little if any undergrowth, but there are a number of vines, lianas, and creepers. Precipitation (the convective type) ranges from 45 to 200 inches (112 to 500 centimeters) annually. This creates soils that are always wet and thoroughly drained of nutrients. Thus, human inhabitants of these regions who clear the land for agriculture must contend with infertile soils.

Tropical deciduous forests are found where precipitation and evaporation engage in seasonal combat, regions that are sometimes wet and sometimes dry. There is too much rainfall for a broad grassland yet not enough for a true rainforest. There are some grassy areas, but trees dominate. Because of the periodic dry seasons, the trees experience dormant periods and so are not evergreen. One of the most commercially usable forests of this type is to be found in Burma, where large stands of teak and mahogany and other important timber products cover much of the land.

Mediterranean woodland and scrub is the typical woodland to be found around the Mediterranean Sea, central Chile, the southern tip of Africa, the southern

Tropical Rainforest in Puerto Rico. This area receives 150 inches (375 centimeters) of rainfall annually.

FIG. 7-2 Earth's Forests.

Tropical rainforest

Tropical deciduous

Mediterranean woodland and scrub

Broadleaf summer deciduous

Midlatitude coniferous forest

Taiga

Mediterranean Woodland and Scrub, Province of Gerona, Spain.

extremities of Australia, and the coastal area of southern California. Plants in these regions include olive, fig, chestnut, and live oak trees, as well as low-growing woody bushes and shrubs, called *maquis* by the peoples of the Mediterranean countries and chaparral by the Californians. Because these are areas of semiaridity (owing to their location on the western sides of the continents, the presence of cool ocean currents that pass offshore, and the dominating influence of the subtropical high pressure belts), the trees and other plants have had to adapt to relatively high temperatures and low precipitation.

Midlatitude coniferous forests contain cone-bearing trees typical of many areas, but nowhere are they so dominant as in the region stretching from the northwestern United States through coastal Canada into Alaska. Here, there are commercially usable stands of redwood, Douglas fir, cedar, and spruce.

Broadleaf summer deciduous forests are located in southern Canada, Europe, China and Japan, southeastern Australia, Argentina, and much of the southern, midwestern, and northeastern sections of the United States. The climate of these areas, which creates long, hot summers, brings forth a wide variety of deciduous trees, hickory, birch, maple, and walnut being the most typical.

Taiga forests (sometimes referred to as **boreal**) are found in the northern two-thirds of Canada in North America and from Sweden in Europe across Russia and Siberia to the Pacific Coast. This vast expanse of conifers is the earth's largest forest. In the northern reaches it is comprised mostly of pine, spruce, and fir, a mixed forest, including birch and maple. The climate is not conducive to human occupancy, which you will see if you contrast the taiga regions with the distribution of population shown in Fig. 1–1. The northern boundary of this region, where the timber resources are virtually untapped, roughly follows the 50°F (10°C) isotherm and blends into the tundra grasslands bordering the Arctic Ocean. There are no trees beyond this

Taiga of Inuvik, Canada's Northwest Territories.

line, for the winters are too long, the summers too short, and there is insufficient precipitation.

GRASSLANDS The earth has many areas where grasses are the predominate form of plant life—the savannas, prairies, steppes, and tundra (Fig. 7–3).

Savanna grasslands might more properly be called tropical grasslands, for they exist roughly from 20° north to 20° south latitudes, and they are included within climatic regions that alternate between wet and dry periods. Trees are present, but usually in small clumps or singly. One might find the umbrella-shaped acacia, palm, and banana trees, as well as shrubs and bushes, especially along streams or around water holes. Savanna grasses are tall and are normally very tough and not well liked by domesticated grazing animals; thus they have not yet been commercially used to any great extent—a fortunate circumstance for the wild herbivores of Africa. The most famous of the earth's savannas are the Sudan, which stretches across Africa south of the Sahara, the llanos of southern Venezuela and eastern Colombia, the campos of south-central Brazil, and the veldt of East and South Africa.

Prairies and **steppes** are regions of the midlatitudes where short grasses are dominant and where trees and large shrubs are found only along streams. Grasses are shorter in steppe regions than in the prairies, but both are quite suitable for grazing

FIG. 7-3 Earth's Grasslands.

209

animals (if the grazing is not overdone). It is interesting to note that the steppe regions of northern Turkestan and the Ukraine of the Soviet Union have been used for pasturage for some 6000 years, and the very rich, black-earth soils have not suffered excessively. Wheat growing was begun centuries ago in the Ukraine with great success, but the conversion of the grazing lands in the eastern steppes to grain lands in 1954 has not fared as well. This is because this region suffers from an unreliable rainfall, periodically creating poor harvests.

Tundra (Russian, for "frozen land") is the name applied to lands bordering the Arctic Ocean, a vast region located north of the 60th parallel that is permanently frozen beneath the surface, in some places down to depths of 700 feet (210 meters). This land of permafrost cannot support large forms of tree life, but a number of plant forms manage to exist in spite of (or because of) the scanty precipitation and the great extremes in temperatures. There are short grasses, sedges (grasslike plants that have stems without joints), mosses, lichens, and tiny 6-inch (15-centimeter) tall willows, which remain so small because of the very short (two months) growing season each year.

The Steppes of the Ukraine: Harvesttime.

Typical Tundra Landscape, Cape Dorset, Canada's Northwest Territories.

DESERTS   The earth's **deserts** (Fig. 7–4) do not support large populations (compare with Fig. 1–1) for the simple reason that most people prefer cooler and more moist climates. In ancient times, the few nomads who managed to exist in these inhospitable regions tended to favor the rain god over the sun god—for good reason. More recently, the discovery of oil reserves in some of the most isolated and desolate spots of the Sahara has brought people into the drilling stations, and with the aid of air conditioners, they are able to survive with some comfort.

The prevailing vegetation forms are called desert shrubs, deciduous plants that are generally leafless until rainshowers force them into immediate flowering, when leaves burst forth and seeds are expelled onto the dry soil. The few evergreen plants have shiny, leathery, and very small leaves and extensive root systems: cactus, Joshua trees (given that name by viewers who were reminded of the Biblical prophet who always spoke with outstretched arms), mesquite (able to survive drifting sand dunes because of its far-reaching branches that manage to stay above the dune surface), sagebrush, and the creosote bush (most common desert shrub of the Sonoran Desert of the United States). After each rainshower, a great variety of wildflowers dots the landscape: daisies, dandelions, and sand verbena.

HIGHLANDS   The vegetation to be found in the higher elevations of the earth, such as the Rocky Mountains of the United States and Canada, the Andes of South America, the Alps of Switzerland, or the Tien Shan of Central Asia, varies considerably with elevation above sea level and latitude. The timber line, for example, is over 10,000 feet (3000 meters) on or near the equator, but it is down to less

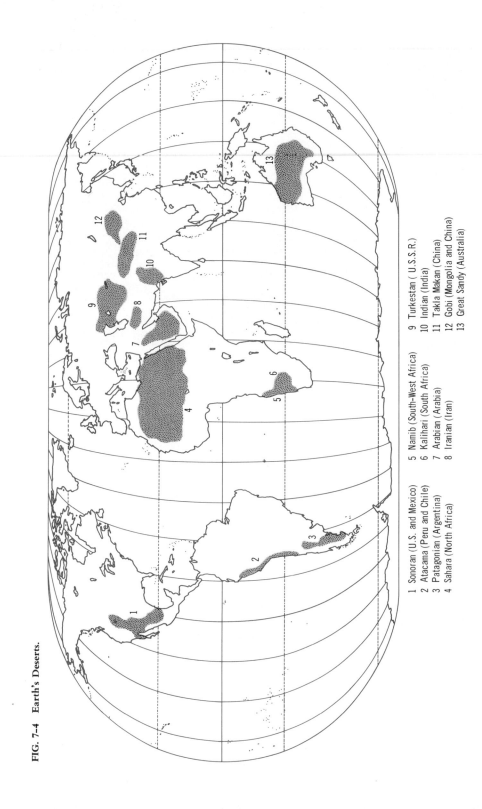

FIG. 7-4 Earth's Deserts.

1 Sonoran (U.S. and Mexico)
2 Atacama (Peru and Chile)
3 Patagonian (Argentina)
4 Sahara (North Africa)

5 Namib (South-West Africa)
6 Kalihari (South Africa)
7 Arabian (Arabia)
8 Iranian (Iran)

9 Turkestan (U.S.S.R.)
10 Indian (India)
11 Takla Makan (China)
12 Gobi (Mongolia and China)
13 Great Sandy (Australia)

than 6000 feet (1800 meters) in the Canadian Rockies. Above this boundary one finds vegetation so similar to that of the tundra regions it has been termed alpine tundra vegetation.

As you move from the foot of the mountains upward, you move from plant community to different plant community, almost as if you were moving from the subtropics northward toward the arctic. For example, let us consider the massive Ruwenzori Mountains of Central Africa. On the west side of these mountains, tropical rainforests approach the foothills, but on the eastern side, the savanna grasses encroach, along with scattered clumps of deciduous trees. If you continue up this eastern slope, you leave the scrub woodland, chop your way through bamboo thickets, hike through pine forests, and eventually, emerge into the tundralike grasses to stare up at the snow-capped peak, devoid of any vegetation.

If, on the way, you were to pass a deep valley, you would notice that the ecosystems change with decreases in elevation down the sides of the valley. These abrupt, lateral changes that occur within short distances make it almost impossible to include the variations in the highland vegetation on a map of the earth's vegetation. The vegetation of vast areas can be averaged out and generalized for most atlases, but the highland regions will be either ignored or "blacked out."

Vegetation and Climate

In our discussion of the earth's weather patterns and climates (chapter 4) we emphasized that weather is a short-term phenomenon and climate is the weather of a region over a long period of time. If you were to travel to a region far from your home, you would know immediately what the weather in the new place was like, but you could not know what climate prevailed unless you had studied the climatic data of the region prior to your arrival or you had knowledge of the vegetation typical of the climate.

Climate exerts a rigorous control over the evolution of any biotic ecosystem, and by carefully examining an area's vegetation you can learn about its climate. Vegetation will tell you about the ratio of precipitation to evaporation, air temperatures, and moisture retentiveness. With this information you might be able to determine climate with a reasonable degree of accuracy. In our next chapter you will learn how Wladimir Köppen used this procedure in 1918 to devise the first acceptable map of the earth's climates.

We might now pose this question: Does vegetation exert any appreciable influence on an area's climate? We know that plants participate in maintaining the hydrologic cycle and the water balance by absorbing moisture from the soil and air and returning it to the air via the process of transpiration. We also know that plants remove carbon dioxide from the air and replace it with oxygen. But the question remains: Does vegetation affect climate, and if so, how much is its influence felt?

The answer is not yet in, although mankind has interfered with the earth's vegetation in many areas, eliminating some climatic climax communities and replacing them with domesticated varieties. We may obtain some of the answers when we have studied the results of wide-ranging plans, such as the one recently introduced by the Russians into western Siberia.

The Soviets have expressed great pride in the vast taiga of that region, claiming that this valuable supply of timber is "inexhaustible." The new plan, however, may not furnish them with the results they expect. While the earlier Soviet plans were not as successful as they had hoped, the plan is to open most of western Siberia, from the Urals eastward to Novosibirsk, to agriculture and livestock production.[4] Will the loss of 25 percent of the taiga affect the climate? Will constant irrigation affect the permafrost condition of the land? Will more or less moisture be available to the atmosphere directly overhead? Will the temperature extremes of Siberia be modified? Will the summertime monsoon-producing Central Asian low-pressure systems be affected?

Review and Discussion

1. Why did people who resided in the world's desert regions differ in their attitude toward the gods and spirits of nature and the sky from people who lived in the colder northern regions?

2. Explain these terms: *epiphytes, thallophytes, hydrophytes.*

3. What is a liana?

4. What is the prime difference between a shrub and a tree?

5. Explain the difference between these terms: *deciduous* and *evergreen.*

6. Explain the term *sclerophyllus.* How does it differ from the term *succulent?*

7. Why do some trees develop needle leaves?

8. What is the main difference in cultural and natural vegetation?

9. Explain the meaning of the term *plant succession.*

10. Explain climatic climax vegetation.

11. What is photosynthesis?

12. What is the effect of high air temperatures on plant growth?

13. What is permafrost?

14. What is meant by *timber line?*

15. Explain the meaning of the term *aspect,* with respect to sunlight.

16. Explain the term *xerophytic.*

17. What type of region would be most apt to have rainforests? Grasslands? Deserts?

18. Explain the differences between savannas, prairies, and steppes.

19. Explain the term *coniferous*.

20. What is the boundary between the tundra and the taiga?

Notes

[1] S. R. Eyre (1963), *Vegetation and Soils: A World Picture* (Chicago, Ill.: Aldine), p. 4.

[2] Eyre, p. 108.

[3] Frits W. Went (1963), *The Plants* (New York: Time–Life Books), p. 82.

[4] Robert G. Kaiser (16 March 1974), "Russian Plans to Open Vast Frontier," *Los Angeles Times.*

Additional Reading

Allen, D. L. (1967). *The Life of Prairies and Plains.* New York: McGraw-Hill.

Cain, S. A. (1944). *Foundations of Plant Geography.* New York: Harper & Row.

Dansereau, Pierre (1957). *Biogeography: An Ecological Perspective.* New York: Ronald Press.

Farb, Peter (1961). *The Forest.* New York: Time–Life Books.

Haden-Guest, Stephen, et al. (1956). *A World Geography of Forest Resources.* New York: Ronald Press.

Leopold, A. S. (1961). *The Desert.* New York: Time–Life Books.

Polunin, Nicholas (1960). *Introduction to Plant Geography.* New York: McGraw-Hill.

U.S. Department of Agriculture (1949). "Trees," *The Yearbook of Agriculture.* Washington, D.C.: U.S. Government Printing Office.

Earth's Climates

Aristotle, the noted Greek philosopher of ancient times, composed a simple classification system for the earth's climates. He came up with three zones: torrid (hot), temperate (mild), and frigid (very cold). Unfortunately, he ignored some important governing influences, such as a place's latitude and altitude, its location relative to oceans or other bodies of water, the dominant vegetation of the region, its position on a land mass, wind systems, and the presence or absence of storms. The student of climate comes soon to the realization that climatic regions are like people: No two can ever be truly identical.

As the preceding chapter on earth's vegetation drew to an end, this question was raised: Does vegetation affect climate, and if so, how much is its influence felt? Another question can be asked: How much influence does climate have on the establishment of the different forms of vegetation? This can be expanded to ask: How important is the relationship of vegetation and climate to mankind?

For more than 2000 years, since Aristotle's attempt to develop a workable classification of climates, scientists of many disciplines have tried to go beyond the attempt to understand why human cultures vary so greatly across the face of the earth. As mankind's knowledge of the earth increased, so did agricultural production and the human population. A British economist, Thomas Robert Malthus, foresaw disaster in that the population was increasing faster than the rate of agricultural production. His dire predictions, first published in 1798, were, fortunately, not fulfilled, although anyone who has studied the increasing population problems of our time realizes there is merit in his theories.

When one considers the spread of people over the earth, the desecration of the soils, vegetation, waters, and air, it becomes apparent that we must do more than merely complain. We must learn more about the intimate relationship between man, the climates, and vegetation, and then actively attempt to find solutions to the problems. We can follow in the footsteps of the botanist who first endeavored to explain the variations of earth's climates by studying the connection between vegetation and climate. This man was Wladimir Köppen.

Köppen's Classification of Climates

As an experienced botanist of the nineteenth century, Köppen was alert to the gradual changes in the vegetative cover from one place to another. As a young man, he studied vegetation maps devised by other botanists and climatologists, who had scientifically examined the changes in vegetation at differing elevations, latitudinal positions, and locations on the European landmass. Although he realized the close relationship of a region's vegetation to its climate, he found himself faced with the same problem every mapmaker faces: How does one pinpoint the exact boundary line between regions of different values? He finally concluded that the line marking the dividing place between two climatic regions was, in reality, a zone of flexibility. He believed this zone was determined by such factors as prevailing seasonal air temperatures, average nighttime and daytime temperatures, the amount and variability of precipitation, the rates of evaporation, and the area's location in relation to large water bodies, winds, ocean currents (cool or warm), and mountainous regions.*

Although Köppen's classification, published in 1918, was greeted with some skepticism in North America, it soon became the standard system. His climatic map (Fig. 8–1) has been modified several times since then, namely by R. Geiger (a German climatologist) and Glenn T. Trewartha (a geographer–climatologist). By comparing this climatic map with Fig. 8–2 the relationship of Köppen's system to the earth's vegetation becomes apparent.

Köppen's classification is based on an alphabetical code (Fig. 8–3). There are five major climate groups:

A   Moist and hot

B   Dry

C   Moist and warm

D   Snowy and cold

E   Polar; icy

In order to differentiate between subdivisions within these major groups, secondary letters were added, which are to be combined to furnish a more complete explanation of the A, C, and D climate regions:

f   Moist all year

m   Monsoon conditions

s   Dry summer, wet winter

w   Dry winter, wet summer

A third letter was added to include information about temperatures:

a   Long, hot summer

b   Short, warm summer

c   Short and moderate
    or cool summer

d   Very cold and dry winter

h   Hot and dry most of
    the year

k   Cold and dry most of the year

The dry climates were assigned the letters W and S to differentiate between desert (BW) and steppe (BS) climate and vegetation conditions. For further explanation, the letters k (cold) and h (hot) are combined with the BW or BS climates.

The polar and subpolar (subarctic) regions begin with the letter E, to which are

*To gain a greater understanding of the problems that confronted Köppen, see: C. W. Thornthwaite (April 1943), "Problems in the Classification of Climates," *The Geographical Review,* 33, no. 2, 233–255. Also available as a Bobbs-Merrill reprint no. G-227.

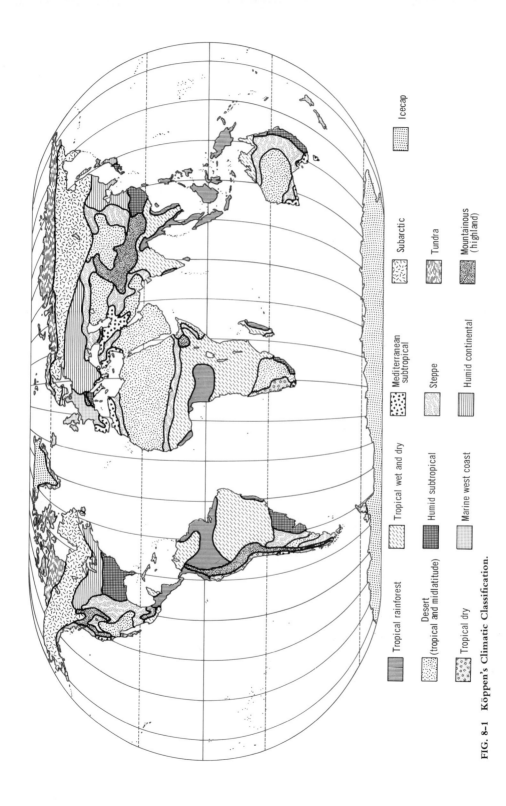

FIG. 8-1 Köppen's Climatic Classification.

Tropical rainforest

Desert (tropical and midlatitude)

Tropical dry

Tropical wet and dry

Humid subtropical

Marine west coast

Mediterranean subtropical

Steppe

Humid continental

Subarctic

Tundra

Mountainous (highland)

Icecap

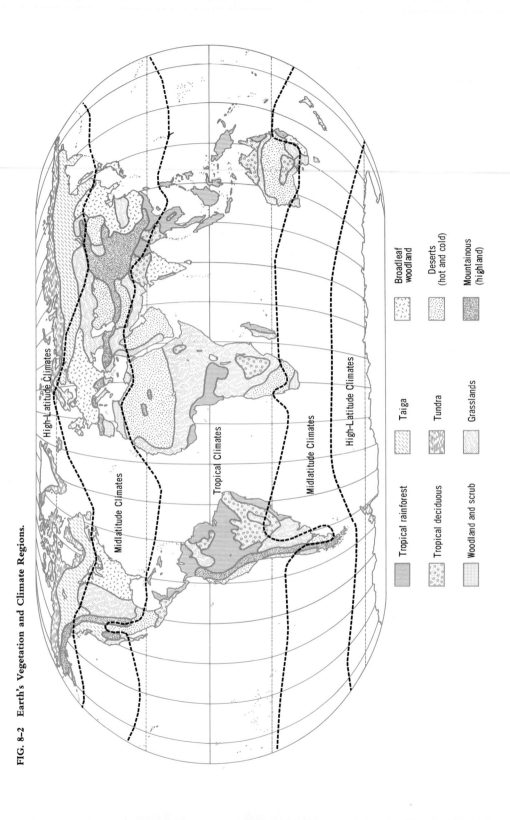

FIG. 8-2 Earth's Vegetation and Climate Regions.

High-Latitude Climates

Midlatitude Climates

Tropical Climates

Midlatitude Climates

High-Latitude Climates

Tropical rainforest

Tropical deciduous

Woodland and scrub

Taiga

Tundra

Grasslands

Broadleaf woodland

Deserts (hot and cold)

Mountainous (highland)

A Tropical or equatorial
(average temperature above 64.4°F(18°C); rainy; no winter)

- *f* Rainy (no dry season)
- *m* Monsoon (excessive rains during certain periods; (dry seasons)
- *w* Wet and dry during winter)

B Dry
(evaporation greater than precipitation)

- *S* Semiarid; steppes or grasslands
  - *h* Dry and hot (average annual temp. over 64.4°F (18°C)
- *W* Deserts;arid
  - *k* Dry and cold (average annual temp. under 64.4°F (18°C)

C Humid, warm, temperate
(seldom below freezing as a monthly average; winter and summer seasons)

- *f* Rainy
  - *a* Long, hot summers
- *s* Summer dry, winter wet
  - *b* Short, warm summers
- *w* Dry season in winter
  - *c* Short, cool summers

D Snowy; subarctic; humid
(northern boundary is the northern limit of forest growth)

- *f* Moist throughout year
  - *a* Long, hot summers
  - *b* Short, warm summers
- *w* Dry season in winter
  - *c* Short, cool summers
  - *d* Very cold winter that is usually dry

E Ice climates
(always cold; warmest month averages under 50°F (10°C)

- *T* Tundra (permafrost and short growing season)
- *F* Icecaps (perpetual ice and snow)

H Highland climates
(mountainous regions with such a variety of weather conditions that climate is impossible to classify)

FIG. 8-3 Köppen's Climate-Classification System.

223

added either *T* (for tundra vegetation and conditions) or *F* (for regions perpetually covered with ice and snow).

Geiger added the letter *H* to account for highland (mountainous) regions that were difficult to include in Köppen's system, and Trewartha added a tertiary code, *n*, for climatic regions plagued by fog. At first glance, this system may seem confusing, but in actual practice it works out well. Table 8–1 provides further clarification. In this table, the climate regions are listed, along with the proper alphabetical codes, plus the typical latitudinal location, dominant air masses, temperatures, annual precipitation, vegetation, soils, and geographic regions representative of the particular climate. To further help you to understand the earth's climates, consult Fig. 8–4, which is a generalized picture of the climatic regions on a hypothetical landmass that extends into both the Northern and Southern hemispheres. Here, you can see the importance of continental position, distance from cool or warm water bodies and ocean currents, and latitude.

Because our concern is with the future of the relationship between humans and the environment, we shall now examine some of the major climate regions to learn what the earth's climates have offered people and how people have responded to this offering.

Tropical Climates

Anthropologists tell us that humans first appeared somewhere within the tropical climate region some 2 or more million years ago, so, perhaps, this is a good place for us to begin our investigation into the influence of climate on humans. Figure 8–5 shows the three major tropical climates, and it is apparent why they are called tropical or equatorial climates: They envelope the equator and extend north and south about 30 to 35 degrees. Because of their latitudinal position and their relationship to the heavy receipt of solar energy throughout the year, temperatures are consistently high.

We do not find any one of the three tropical climates dominating over these regions because of the differences in continental position, the presence of cold or warm ocean currents, and the wind systems (compare with Figs. 4–7 and 4–8). While humans may have originated somewhere within this equatorial region, they apparently had little desire to remain there. If you consult the map of world population distribution (Fig. 1–1), you will note that, while there are some tropical regions with relatively high population density, most are sparsely populated.

TROPICAL RAINFOREST   The **tropical rainforest** climate region is often referred to as equatorial rainforest or tropical rainy, and the letters *Af* are used for its classification code. The largest expanses of this climate will be found within 10 degrees north and south of the equator. The average daily temperatures range between 80°F (27°C) and 90°F (32°C) (Fig. 8–6) and seldom fall below 64.4°F (18°C) at night. Because this region is a part of the equatorial low-pressure depression, the convectional uplift of the moist, heated air creates daily thunderstorms. The average annual precipitation ranges from 65 to 100 inches (162 to 250 centimeters), and there is

FIG. 8-4 Earth's Climates (Generalized).

225

TABLE 8-1 Climatic regions of the world

CLIMATE REGION	CLIMATIC SYMBOL	LATITUDINAL POSITION	AIR MASS	TEMPERATURE	PRECIPITATION ANNUALLY	VEGETATION	SOILS	TYPICAL REGIONS
Tropical rainforest	Af Am	10°N–10°S to 20°N	mT mT	Day: 90°–95°F (32°–35°C) Night: 70°F (21°C)	65–100 in. (162–250 cm)	Selva: broad-leaf, ever-green trees	Latosols	Amazon, Meso-America, Guinea coast, Congo, Ceylon, SE Asia
Tropical dry	BWh	15°–35°N,S	cT	Above 32°F (0°C)	Less than 10 in. (25 cm)	Xerophytic shrubs	Sierozems	Sahara, Arabian, Thar, Australian, Sonoran deserts; Kalihari, Atacama
Tropical wet and dry	Aw	5°–25°N,S	mT	Winter: 70°F (21°C) Summer: 80°–90°F (27°–32°C)	Summer rainy	Tropical savanna	Latosols	Sudan, veldt, llanos
Mediterranean subtropical	Csa	30°–45°N,S West coasts	mP (summer) mT (winter)	Summers: 70°F (21°C) Winters: 50°F (10°C)	15–20 in. (37–50 cm)	Chaparral, maquis: sclerophyll	Reddish chestnut; reddish brown	Mediterranean region; southern California; central Chile; tip of South Africa
Humid subtropical	Cfa	20°–35°N,S East coasts	mT	Warm months average above 71.6°F (22°C)	40 in. (100 cm)	Broad leaf, mixed broad leaf	Podzols	SE U.S.A.; SE China; southern Japan; eastern Australia
Marine west coast	Cfb	40°–60°N,S West coasts	mP	Warmest month average below 71.6°F (22°C)	23 in. (57.5 cm) low 200 in. (500 cm) maximum	Coniferous forest, broad leaf	Podzols	NW U.S.A.; Europe; southern Chile

TABLE 8-1 (continued)

CLIMATE REGION	CLIMATIC SYMBOL	LATITUDINAL POSITION	AIR MASS	TEMPERATURE	PRECIPITATION ANNUALLY	VEGETATION	SOILS	TYPICAL REGIONS
Midlatitude dry (steppe, desert)	BSk BWk	35°–50°N,S	cT (summer) cP (winter)	Less than 64.4°F (18°C)	10–20 in. (25–50 cm)	Short-grass prairie	Brown; reddish	Patagonia, Gobi, Takla Makan, western Great Plains
Humid continental	Dfa	35°–60° North only	cP (winter) mT (summer)	32°F (winter) (0°C) 72°F (summer) (22°C)	30–50 in. (75–125 cm)	Mixed hardwoods	Gray-brown podzols	NE U.S.A. and southern Canada, upper Midwest, Soviet Union, eastern China
Continental subarctic (boreal)	Dfc	50°–70°N	mP (summer) cP (winter)	Less than 4 months with average above 50°F (10°C)	10–25 in. (25–63 cm)	Taiga: needleleaf evergreen	Permafrost	Alaska, northern Canada, northern U.S.S.R. and Siberia
Tundra	E, ET	60°–75°N	mP	32°–50°F (0°–10°C)	Less than 10 in. (25 cm)	Sedges, lichens, shallowroot grasses	Lithosols	Northern ·portion North America, Asia
Polar icecap	EF	Icecaps at both poles	Arctic; Antarctic	No above-freezing month average	Less than 10 in. (25 cm)	None	None	Greenland, northernmost tip of Antarctica
Highland	H	At all latitudes	Too variable to classify	Large diurnal range	Orographic	Too variable to classify	Lithosols on ridges and high valleys; alluvium in low valleys	Mountainous areas of world

FIG. 8-5   Tropical Climates of the World.

Labrador Current

California Current

Canary Current

Peru (Humboldt) Current

Benguela Current

Australian Current

West Wind Drift

Tropical rainforest (*Af, Am*)

Tropical wet and dry (*Aw*)

Tropical dry (*BWh*)

Warm ocean currents

Cold ocean currents

no dry season. Someone once said that in these tropical rainforest areas there are only wet and less wet seasons.

The average daily pattern is for cumulus clouds gradually to form as the land heats up. By early afternoon the clouds become cumulonimbus (rainbearing) and begin precipitating as they move slowly westward, pushed by the trade winds. Should a mountain stand in the way, the precipitation will be heavier—as we saw in chapter 6 when we considered the rainfall for Mt. Waialeale in Hawaii, which averages 471.66 inches (1180 centimeters) of rainfall annually. Such areas of extremely heavy precipitation occur along windward coasts of landmasses, where warm ocean currents make available a plentiful supply of warm and moist air masses.

While the tropical rainforest regions are not attractive to people for permanent residence, they have supplied us with valuable products such as rubber, cocoa, medicines, and timber (mahogany, teak, ebony). The constant rainfall also exerts a tremendous influence on the soils, washing away the dissolvable minerals and the little humus that does manage to escape the increased activity of bacteria. The resulting soils, called **latosols** (from the Latin *later,* "brick") are able to support the luxuriant stands of trees, but agriculturists (both primitive and advanced) have found that after the forest is cleared, the soils lose their fertility within a year or two and fertilizers become a necessity.

TROPICAL MONSOON The **tropical monsoon** climate region (*Am*) appears to be concentrated in Southeast Asia (Bangladesh, Burma, and Malaysia), the East Indies (Sumatra, Java, New Guinea, and Celebes), the Philippines, and the area near the mouth of the Amazon River of Brazil. Monsoon climate regions have heavy rainfall, caused by monsoon conditions (see Fig. 4–12) during the summer periods. Intense heating of the interior of the landmasses draws in moist air from the nearby oceans, and the resulting rains often cause serious damage. Cherapunji (in the Indian state of Assam) once recorded 1041.87 inches (2605 centimeters) of rain in the period from August 1860 to July 1861. So great is this problem to the people of Bangladesh that, in the area most often hit by monsoon rains, the Bengalis have begun building man-made hills to serve as a place of refuge when the waters rise too high. Not all monsoons are killers, however, for they do bring water to a normally parched land every summer. Without this rain, the millions of people who must exist on extremely low-nourishment diets could not survive.

TROPICAL WET AND DRY Surrounding the rainforest climates are the great expanses of savanna grasslands, a climate region known as **tropical wet and dry**(*Aw*). Here are the tall-grass areas, such as the Sudan (south of the Sahara Desert), the veldt (East and South Africa), the campos (Brazil), the llanos (Venezuela), and the savannas of India, Southeast Asia, and northern Australia. Savannas receive their rain during the summer, as the nearby oceans are heated and tropical maritime air masses are dominant. The winters are dry, with average precipitation of less than 2 inches (5 centimeters). The climograph for Fig. 8–7 shows that Darwin, Australia, averages about 83°F (28°C) for the year, with the yearly rainfall amounting to about 60 inches (155 centimeters). But, see how this precipitation is spread out through the

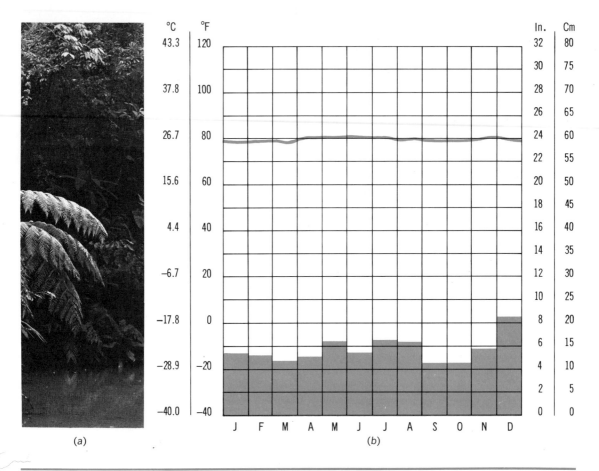

		°C	°F			In.	Cm

(a)

(b)

CLIMATE: Af	STATION: Changuinola, Panama	LAT.: 9°29′N	LONG.: 82°29′W	ELEVATION: 53 feet (16 meters)

	Jan.	Feb.	Mar.	Apr.	May	June	July	Aug.	Sept.	Oct.	Nov.	Dec.	Year
Temperature													
°F	79.3	78.6	79.5	80.4	81.3	81.7	80.4	81.0	81.1	81.3	80.1	79.0	80.2
°C	26.3	25.9	26.4	26.9	27.4	27.6	26.9	27.0	27.3	27.4	26.7	26.1	26.8
Precipitation													
Inches	8.9	7.5	6.1	6.9	10.2	7.9	11.1	10.4	5.2	5.3	9.2	16.1	104.7
Centimeters	22.3	18.8	15.3	17.3	25.5	19.8	27.8	26.0	13.0	13.3	23.0	40.3	261.8

(c)

FIG. 8-6   Tropical Rainforest (*Af*): Changuinola, Republic of Panama.   The dense foliage and giant ferns that line this river in the interior are the result of high monthly temperature averages and heavy precipitation. Climatic data can be presented in any one of three different ways: (*a*) photograph; (*b*) climograph, a visual display of the data included in the climatic chart (*c*). (Data taken from U.S. Department of Commerce, *World Weather Records, 1951–60,* 1965.)

CLIMATE: Aw		STATION: Port Darwin, Australia			LAT.: 12°26'S		LONG.: 130°52'E		ELEVATION: 89 feet (27 meters)				
	Jan.	Feb.	Mar.	Apr.	May	June	July	Aug.	Sept.	Oct.	Nov.	Dec.	Year
Temperature													
°F	82.2	81.7	82.9	82.8	81.0	77.7	76.8	78.4	81.9	84.4	84.6	83.8	81.5
°C	27.9	27.6	28.3	28.2	27.0	25.4	24.9	25.8	27.7	29.1	29.2	28.8	27.5
Precipitation													
Inches	15.0	13.7	9.8	6.3	0.68	0.04	0.11	0.02	0.41	3.09	6.02	9.5	64.6
Centimeters	37.5	34.3	24.5	15.8	1.7	0.1	0.28	0.05	1.03	7.73	15.1	23.8	161.5

FIG. 8-7 Tropical Wet and Dry (Aw): Port Darwin, Australia. (Data taken from U.S. Department of Commerce, *World Weather Records, 1951–60, 1965.*)

year. The winter months (May to September) average only 0.28 inches (0.7 centimeters), and the summer months (October to April) average almost 8.5 inches (21 centimeters). Soils in tropical wet and dry regions are not considered to be very fertile, and the grasses seem to be too tough and tall for successful programs of animal husbandry (sheep and cattle raising), so people are discouraged from migrating to these areas. It might be noted, however, that African wildlife seems to flourish in these regions.

TROPICAL DRY   The tropical dry climate regions are called **desert** (*BW*) and extend from 15° to 35° north and south of the equator. They surround the tropical wet and dry climate regions. These deserts are under the dominating influence of the subtropical high-pressure areas of descending air masses, which bring little precipitation. In regions such as North Africa's Sahara, Australia's Great Sandy, India's Thar, and the Arabian Desert, evaporation usually exceeds precipitation, and the average annual precipitation is less than 10 inches (25 centimeters). The tropical dry regions always have a west-coast frontage, extending eastward in some cases for great distances. Some deserts, such as the Atacama of Peru and Chile, are narrow strips bounded on the east by high mountain ranges. The characteristics of the tropical dry regions are also affected by the presence nearby of cool ocean currents, which are subject to less evaporation. Westerly winds blowing across these cool waters are not able to receive or retain much moisture, much less transport it over the parched lands beneath them.

Figure 8–8 tells you that desert lands suffer considerably from the combination of warm temperatures and low precipitation. Cairo, Egypt, has an average yearly temperature of about 71°F (22°C) and a yearly total of only 1.3 inches (3.25 centimeters) of rain. Summer months show a zero amount, and the total for the "rainy" winter comes to only 1.3 inches (3.25 centimeters). Desert regions have become noted for setting records for extremes in the temperature ranges. The official world's record of 136.4°F (58°C) as the highest temperature ever recorded was set at El Aziza, Libya, a city located about 20 miles (32 kilometers) inland from

CLIMATE: BWh		STATION: Beni Abbes, Algeria			LAT.: 30°08′N		LONG.: 02°10′W		ELEVATION: 1643 feet (501 meters)				
	Jan.	Feb.	Mar.	Apr.	May	June	July	Aug.	Sept.	Oct.	Nov.	Dec.	Year
Temperature													
°F	51.4	56.5	64.4	72.1	80.1	89.96	96.4	94.1	86.2	73.4	61.9	52.7	73.2
°C	10.8	13.6	18.0	22.3	26.7	32.2	35.8	34.5	30.1	23.0	16.6	11.5	22.9
Precipitation													
Inches	3.73	0.09	0.27	0.32	0.04	0.001	0.02	0.08	0.12	0.41	0.26	0.49	2.47
Centimeters	9.3	22.5	0.68	0.8	0.1	0.003	0.05	0.2	0.3	1.03	0.65	1.23	6.18

FIG. 8-8 Tropical Dry (*BWh*): Oasis of Timimoun near Beni Abbes, Algeria. The oasis is located in the Grand Erg Occidental (Great Western Sandy Desert). (Data taken from U.S. Department of Commerce, *World Weather Records, 1951–60,* 1965.)

the Mediterranean Sea. The record for the greatest diurnal (daily) range was set at In Salah, midway between the Atlas and Ahagger mountains of the Sahara, when, during one 24-hour period the temperature ranged from 26°F (−3°C) at night to 126°F (52°C) the next day—a range of 100°F (57°C).

Subtropical Climates

The earth's subtropical climates are situated on the poleward sides of the tropical climates, and in these climate regions some of the world's heaviest concentrations of population are to be found (Fig. 8–9). There are two subtropical climates, the west coastal dry subtropical, also known as Mediterranean subtropical, and the east coastal humid subtropical, also called humid subtropical.

Midlatitude Climates

The climate regions situated in the middle latitudes, between 30° and 60° north latitude and 30° to 50° south latitude, are generally temperate climates with high precipitation, except for the interior regions, which are quite dry. Figure 8–12 maps the locations of the three midlatitude climate regions: marine west coast, humid continental, and midlatitude dry.

MARINE WEST COAST The term *west coast* does not apply to **marine west coast** climate (*Cfb*) regions in South Africa or Australia, but only to southern Chile in South America, a narrow strip along the western North American coast from the state of Washington on through British Columbia to the southern tip of Alaska, and most of northwestern Europe. This climate region normally receives considerable

FIG. 8-10 West Coastal Dry Subtropical (*Csa*): Wine Country North of Rome, Italy. (Data taken from U.S. Department of Commerce, *World Weather Records, 1951–60,* 1965.)

CLIMATE: Csa	Jan.	Feb.	Mar.	Apr.	May	June	July	Aug.	Sept.	Oct.	Nov.	Dec.	Year
STATION: Rome, Italy LAT.: 41°48′N LONG.: 12°36′E ELEVATION: 432 feet (131 meters)													
Temperature													
°F	45.5	47.1	50.7	55.9	63.3	70.9	75.4	75.4	69.9	61.2	53.4	48.6	59.7
°C	7.5	8.4	10.4	13.3	17.4	21.6	24.1	24.1	21.1	16.2	11.9	9.2	15.4
Precipitation													
Inches	2.56	2.5	2.2	2.56	2.03	1.16	0.98	1.07	2.61	3.07	3.79	3.8	28.4
Centimeters	6.4	6.25	5.5	6.4	5.08	2.9	2.45	2.68	6.53	7.68	9.48	9.5	71.0

precipitation, as Fig. 8–13 shows. Annual precipitation in the Vancouver area ranges from a low of 23 inches (58 centimeters) to an occasional high of around 80 inches (200 centimeters). Yearly temperature averages are about 48°F (9°C), and the amount of precipitation is heaviest during the winter months, totaling 54 inches (135 centimeters) from September through May, and averaging 6 inches (15 centimeters) per month. This results from the generation of low-pressure systems that move out of the polar front off Alaska. Windward slopes of the coastal mountain ranges receive considerably more annual precipitation than low-lying areas such as Vancouver, as is evidenced by the record set at the Olympic Mountains of Washington—an average annual precipitation of 150.73 inches (377 centimeters).

One might think that the general climate of this region would be rather inhospitable, but such is not the case because of the proximity of the ocean. The oceans

FIG. 8–11 Humid Subtropical (*Cfa*): Hurricane Mills, Southwest of Nashville, Tennessee. (Data taken from U.S. Department of Commerce, *World Weather Records, 1951–60*, 1965.)

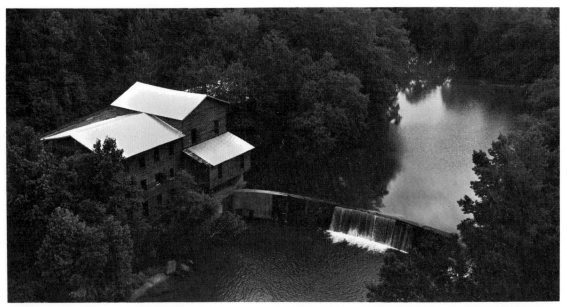

CLIMATE: *Cfa*	STATION: *Nashville, Tennessee*		LAT.: *36°07′N*		LONG.: *86°41′W*		ELEVATION: *584 feet (178 meters)*					
	Jan.	Feb.	Mar.	Apr.	May	June	July	Aug.	Sept.	Oct.	Nov.	Dec.	Year
Temperature													
°F	39.9	42.1	49.1	59.5	68.5	77.4	80.2	79.2	72.9	61.5	48.6	41.4	60.1
°C	4.4	5.6	9.5	15.3	20.3	25.2	26.8	26.2	22.7	16.4	9.2	5.2	15.6
Precipitation													
Inches	5.5	4.6	5.2	3.74	3.7	3.27	3.7	2.87	2.87	2.32	3.27	4.17	45.9
Centimeters	13.8	11.5	13.0	9.34	9.25	8.18	9.25	7.18	7.18	5.8	8.18	10.4	114.8

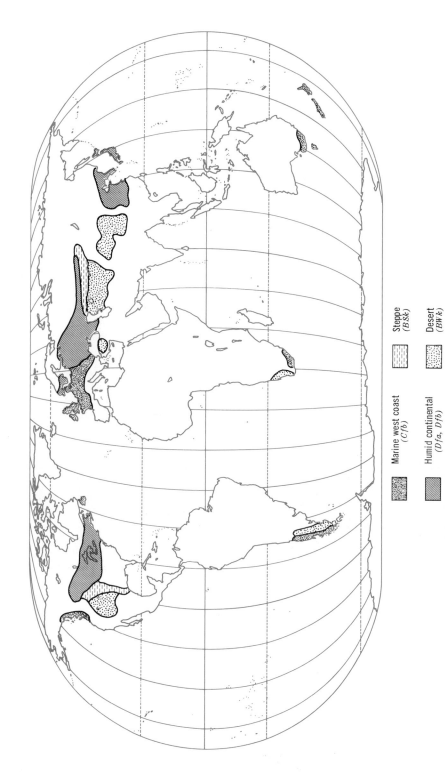

Steppe
(BSk)

Desert
(BWk)

Marine west coast
(Cfb)

Humid continental
(Dfa, Dfb)

FIG. 8-12 Midlatitude Climates of the World.

237

exert a valuable moderating influence on the land, keeping down any extremes in temperature so that the range between the warmest and coldest months averages about 27°F (15°C). This may be a region of hunters and fishermen, but farmers can have a profitable agricultural business raising fruits and vegetables during the relatively long growing season (approaching 200 days). In addition, the great expanse of coniferous and broadleaf forests supplies timber.

HUMID CONTINENTAL   The cold and snowy **humid continental** climates (*Dfa*, *Dfb*) are to be found only in the Northern Hemisphere, extending from 35° to 60° north latitude. In North America the western boundary of this climate region is located in the center of the continent, and from there it stretches to the Atlantic coast, taking in the southern one-third of Canada and the upper Midwest of the United

FIG. 8-13   Marine West Coast (*Cfb*): Kitsilano Park, Vancouver, British Columbia, Canada.   During the summer, sunbathers enjoy the cooling waters off Vancouver, the skyline of the city, and the mountains of the Coast Range in the background. (Data taken from U.S. Department of Commerce, *World Weather Records, 1951–60, 1965.*)

CLIMATE: *Cfb*		STATION: *Vancouver, Canada*			LAT.: 49°11′N		LONG.: 123°10′W			ELEVATION: *17 feet (5 meters)*		
	Jan.	Feb.	Mar.	Apr.	May	June	July	Aug.	Sept.	Oct.	Nov.	Dec.	Year
Temperature													
°F	36.1	39.6	42.4	48.4	54.7	59.4	63.7	62.6	57.7	50.2	42.8	39.0	49.7
°C	2.3	4.2	5.8	9.1	12.6	15.2	17.6	17.0	14.3	10.1	6.0	3.9	9.8
Precipitation													
Inches	5.56	4.84	3.78	2.36	1.89	2.01	1.02	1.42	2.2	4.68	5.68	6.24	41.9
Centimeters	13.9	12.1	9.45	5.9	4.73	5.03	2.55	3.55	5.5	11.7	14.2	15.6	104.8

States. As for Europe and Asia, the humid continental climate is found in western Siberia, much of Russia and the Ukraine, and the southern portions of Finland and Sweden. Like many other climate regions, this climate witnesses the battles issuing from the polar front during the winter and from the movement northward of tropical maritime air masses during the summer months.

Winters can be very cold, as shown in Fig. 8–14. The average annual temperature is a low 45°F (7°C), ranging from a January low of 22°F (−5.5°C) to a July high of 69°F (20.5°C). Snowy winters are common and often encompass 9 or 10 months of a year, especially in the northern sections, which have cooler summers (*Dfb*). Incidentally, the northern border of the *Dfb* climate is the northernmost limit for conventional forms of agriculture. Precipitation totals about 32.5 inches (81 centimeters) for the year, with a monthly average of 2.7 inches (6.75 centimeters). While these conditions create problems for farmers, they are advantageous for timbermen, because the humid continental regions are noted for the mixed forests of hardwoods and softwoods, broadleaf deciduous and needle-leafed evergreens.

The southern portion of the humid continental climate region (*Dfa*) receives heavier amounts of precipitation, with monthly averages advancing to the 3- and 4-inch (7- and 10-centimeter) levels. Summer rainfall accompanying tropical cyclonic storms can be quite severe, as is evidenced by the 31-inch (78-centimeter) downpour experienced at Smethport, Pennsylvania, on July 18, 1942, during a 4½-hour period.

MIDLATITUDE DRY   The drier regions of the middle latitudes (*BSk, BWk*), called **midlatitude dry** climates, are similar in many respects to those of the lower latitudes. Grasses and desert shrub are the dominant vegetation types, along with scattered clumps of broadleaf deciduous trees. The prime difference is the temperature. Tropical deserts, for example, have monthly temperatures that remain above the point of freezing, while midlatitude deserts often have one or two months when the temperature falls below 32°F (0°C). The reason is the latitudinal position (relative to the angle of insolation) and the presence of high-pressure systems over Canada and Siberia. Air drawn from the colder northern regions is forced down into the deserts during the winters, causing a steep decline in temperature (Fig. 8–15). The yearly tempera-

FIG. 8–14   Humid Continental with Cool Summer (*Dfb*): Toronto, Ontario, Canada.   (Data taken from U.S. Department of Commerce, *World Weather Records, 1951–60,* 1965.)

CLIMATE: *Dfb*	STATION: *Toronto, Canada*		LAT.: 43°40′N		LONG.: 79°24′W			ELEVATION: 383 *feet (117 meters)*				
	Jan.	Feb.	Mar.	Apr.	May	June	July	Aug.	Sept.	Oct.	Nov.	Dec.	Year
Temperature													
°F	38.8	38.6	32.3	44.4	55.8	66.2	71.2	69.6	61.5	51.1	39.7	28.9	47.7
°C	3.8	3.7	0.2	6.9	13.2	19.0	21.8	20.9	16.4	10.6	4.3	−1.7	8.7
Precipitation													
Inches	2.64	1.93	2.64	2.6	2.76	2.48	2.91	2.4	2.56	2.36	2.48	2.4	31.0
Centimeters	6.6	4.8	6.6	6.5	6.9	6.2	5.4	6.0	6.4	5.9	6.2	6.0	77.6

ture average for Tashkent, center of the cotton-growing region of Soviet Central Asia, is a low 56°F (13.3°C), while the summer months see a temperature average of around 80°F (27°C).

The midlatitude steppes and prairies (*BSk*) are sometimes barren appearing, yet these short-grass regions are vitally important. If you will again look at Fig. 7–3, you will note the areas made famous as grazing lands for beef cattle: the Great Plains of the United States, the pampa of Argentina, and the Kirghiz steppes of Central Asia and the Ukraine (Fig. 8–16). The veldt of South Africa, notably the steppes of the Kalihari, is now becoming more extensively used for pasturage. The Australians are justifiably proud of the grassy "outback" region that lies west of the Eastern Highlands.

These interior regions are cut off from moisture-laden warm air masses; thus they remain cool throughout the year. Consider also the situation for Cheyenne, Wyoming, where the yearly temperature average reaches 44.5°F (6.9°C), and at least three months of the year have average temperatures below freezing. As for precipitation, the annual total of less than 15 inches (38 centimeters) is an average of slightly more than 1 inch (2.5 centimeters) per month. The chinook winds that sweep down the rain-shadow side of the Rockies during the winter months are drying winds, devoid of moisture. Maritime tropical air masses sweep northward from the Gulf of Mexico and the Caribbean during the summer months, and as they move over the heated land, convectional precipitation results. The four summer months account for 7.7 inches (19 centimeters) of the yearly total, while the remaining eight months account for only 6.7 inches (17 centimeters).

High-Latitude Climates

The high-latitude climatic regions (Fig. 8–17) are to be found poleward of the 50° parallel and include the continental subarctic (*Dfc*, *Dwd*, and *Dwc*), tundra (*ET*), and the icecaps (*EF*).

FIG. 8–15 Midlatitude Desert with Cool Summer (*BWk*): Anhsi, China. This station is situated on the eastern margin of the Tarim Basin, close to the Gobi Desert. (Data taken from U.S. Department of Commerce, *World Weather Records, 1951–60,* 1965.)

CLIMATE: BWk	STATION: *Anhsi, China*		LAT.: 40°43′N		LONG.: 95°57′E			ELEVATION: *3900 feet (1189 meters)*					
	Jan.	Feb.	Mar.	Apr.	May	June	July	Aug.	Sept.	Oct.	Nov.	Dec.	Year
Temperature													
°F	17.6	26.0	40.6	55.6	66.4	75.2	78.8	76.5	65.1	50.5	32.0	16.9	50.2
°C	−8.0	−3.3	4.8	13.1	19.1	24.0	26.0	24.7	18.4	10.3	0.0	−8.4	10.1
Precipitation													
Inches	0.08	0.04	0.16	0.22	0.12	0.21	0.63	0.44	0.1	0.04	0.06	0.19	2.28
Centimeters	0.2	0.1	0.4	0.55	0.3	0.53	1.58	1.1	0.25	0.1	0.15	0.48	5.7

CONTINENTAL SUBARCTIC The **continental subarctic,** or **boreal,** climate (*Dfc, Dwc, Dwd*) region is the most important commercially of the three high-latitude climates, primarily for the vast expanse of timber available from the taiga, a mixed forest of hardwoods. This very cold region is dominated by polar air masses throughout the year and is constantly being swept by cyclonic storms. Traditional forms of agriculture are not practiced here. Figure 8–18 details conditions in Irkutsk, Siberia. This city is situated about 36 miles (40 kilometers) from Lake Baikal and is a center for the timber industry of the region. The average temperature for the year approaches 31°F (−0.6°C), but the short summer sometimes attains tempera-

FIG. 8-16 Midlatitude Steppe (*BSk*): Rostov-on-Don, Ukraine S.S.R. (Data taken from U.S. Department of Commerce, *World Weather Records, 1951–60,* 1965.)

CLIMATE: *BSk* STATION: *Rostov-on-Don, Ukraine S.S.R.* LAT.: 47°15′N LONG.: 39°40′E ELEVATION: *254 feet (77 meters)*												
	Jan.	Feb.	Mar.	Apr.	May	June	July	Aug.	Sept.	Oct.	Nov.	Dec.	Year
Temperature													
°F	22.5	23.2	31.8	48.9	62.2	69.6	74.3	72.1	61.5	48.2	36.3	27.1	48.2
°C	−5.3	−4.9	−0.1	9.4	16.8	20.9	23.5	22.3	16.4	9.0	2.4	−2.7	9.0
Precipitation													
Inches	1.52	1.64	1.28	1.56	1.44	2.32	1.96	1.48	1.28	1.76	1.6	1.48	19.3
Centimeters	3.8	4.1	3.2	3.9	3.6	5.8	4.9	3.7	3.2	4.4	4.0	3.7	48.3

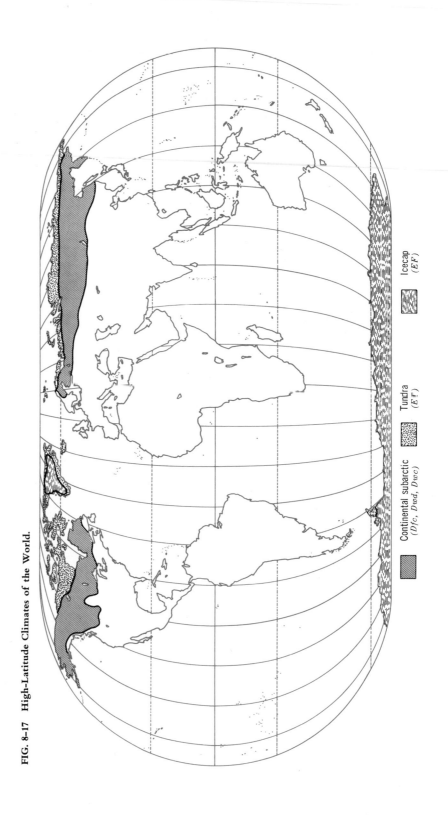

FIG. 8-17 High-Latitude Climates of the World.

Continental subarctic
(*Dfc, Dwd, Dwc*)

Tundra
(*ET*)

Icecap
(*EF*)

tures in the low 70s F (20s C). Precipitation is concentrated in the summer months (almost two-thirds of the total).

TUNDRA North of the boreal forest, or taiga, is the usually drab region called the **tundra** (*ET*). Here, maritime polar air masses are dominant throughout the year. Precipitation (see Fig. 8–19) in Pt. Barrow, Alaska, amounts to about 5 inches (12.5 centimeters) per year and is concentrated in the period of late fall. Temperatures are below freezing for most of the year, with only the months of July and August creeping over that point. Although the ground is permanently frozen to depths of hundreds of feet in most places, there is vegetation to be seen: sedges, mosses, lichens, and tiny, dwarf willow trees. The vegetation springs to life during the short summer and presents a dazzling display of colorful flowers.

The permafrost presents problems to the tundra's human inhabitants, especially if they construct houses or buildings that are large and without adequate insulation in or under the floors. A visitor may be surprised to see a house tilted from the horizontal, one side having sunk down into the upper 10 inches (25 centimeters) of thawed soil! If a building were constructed on a steep slope in this region, it would have to be firmly anchored deep in the permafrost or it would slide down the hill when the summer thaw arrived.

ICECAP The **icecap** climate (*EF*) region embraces most of Greenland, the floating ice pack of the Arctic, and the great continental glacier that covers the continent of Antarctica. There is no true soil in this region, for the temperatures are too cold, and there is no bacterial activity.

The tremendous amount of water encased in these ice regions has an important bearing on our supply of water and on the temperature of air masses that move out to other regions of the earth. Antarctica contains about 8 times as much ice as all of the Arctic Ocean and Greenland combined, and the ice is 1 mile (1.6 kilometers) thick in many places. Consulting Fig. 8–20, you may wonder how the icecaps can build to such depths when there is so little precipitation—less than 5 inches (12.5 centimeters) per year. The answer is found in the high albedo (reflectiveness) of

FIG. 8-18 Continental Subarctic (*Dwc*): Irkutsk, Siberia. This station is the center for the Russian timber industry. (Data taken from U.S. Department of Commerce, *World Weather Records, 1951–60*, 1965.)

CLIMATE: Dwc	STATION: Irkutsk, U.S.S.R.			LAT.: 52°16′N		LONG.: 104°21′E			ELEVATION: 1930 feet (588 meters)				
	Jan.	Feb.	Mar.	Apr.	May	June	July	Aug.	Sept.	Oct.	Nov.	Dec.	Year
Temperature													
°F	−5.44	−0.04	15.26	34.9	47.8	59.7	64.2	59.2	46.8	33.9	12.56	−1.3	30.6
°C	−20.8	−17.8	−9.3	1.6	8.8	15.4	17.9	15.1	8.2	1.1	−10.8	−18.5	−0.8
Precipitation													
Inches	0.48	0.32	0.36	0.6	1.16	3.32	4.08	3.96	1.96	0.8	0.68	0.6	18.3
Centimeters	1.2	0.8	0.9	1.5	2.9	8.3	10.2	9.9	4.9	2.0	1.7	1.5	45.8

CLIMATE: ET	STATION: Pt. Barrow, Alaska		LAT.: 71°18′N		LONG.: 156°47′W			ELEVATION: 30 feet (9 meters)				
	Jan.	Feb.	Mar.	Apr.	May	June	July	Aug.	Sept.	Oct.	Nov.	Dec.	Year
Temperature													
°F	−16.2	−18.2	−14.6	0.14	18.3	33.1	39.0	37.9	30.6	16.5	−0.76	−11.2	9.68
°C	−26.8	−27.9	−25.9	−17.7	−7.6	−0.6	3.9	3.3	−0.8	−8.6	−18.2	−24.0	−12.4
Precipitation													
Inches	0.2	0.16	0.12	0.12	0.12	0.36	0.8	0.92	0.64	0.52	0.24	0.16	4.4
Centimeters	0.5	0.4	0.32	0.3	0.3	0.9	2.0	2.3	1.6	1.3	0.6	0.4	11.0

FIG. 8-19 Tundra (ET): Pt. Barrow, Alaska. This area is noted for its permafrost, which thaws at its surface each spring, creating numerous rivers and lakes that are often filled with floating ice. (Data taken from U.S. Department of Commerce, *World Weather Records, 1951–60*, 1965.)

the ice fields and in the low evaporation. What precipitation an icecap receives, it retains (except when the outer sections break off and float away on the river or ocean current).

Highland Climates

For **highland** climate regions (*H*), we must make the same statements we made when discussing the vegetation of mountainous areas: There are too many variables to permit generalizations. The climate is influenced by such determinants as whether there are high or small mountains, broad ranges or narrow ones, deep and wide valleys or shallow and narrow ones, whether or not there are interior plateaus. In addition, of course, an important variable is a region's latitude.

Climate and Mankind

In our discussions this far, we have attempted to understand the relationship between vegetation and climate, and how this relationship affects humans. We have seen how people migrate to areas that have the "best" climates, those most comfortable and suitable for agriculture. We have also touched lightly on the problem of the expanding human population and the ever-present threat of overpopulation and sufficient means of subsistence. While there are many who insist that human ingenuity will discover solutions to the problems of energy shortages, pollution of soils, air, and water; and the reduction of the world's forests, there is also the danger that climatic

FIG. 8-20 Icecap (EF): McMurdo Sound, Antarctica. This station is in a land of perpetual ice and snow. (Data taken from U.S. Department of Commerce, *World Weather Records, 1951–60,* 1965.)

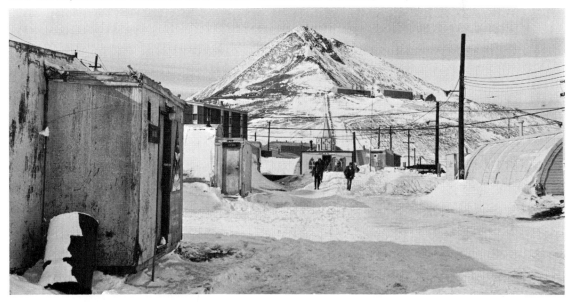

CLIMATE: *EF*		STATION: *McMurdo Sound, Antarctica*			LAT.: 77°50′S		LONG.: 166°30′E		ELEVATION: *150 feet (46 meters)*				
	Jan.	Feb.	Mar.	Apr.	May	June	July	Aug.	Sept.	Oct.	Nov.	Dec.	Year
Temperature													
°F	25.5	16.3	−1.12	−8.5	−9.58	−12.3	−16.9	−20.2	−10.1	−4.72	14.7	25.3	−0.22
°C	−3.6	−8.7	−18.4	−22.5	−23.1	−24.6	−27.2	−29.0	−23.4	−20.4	−9.6	−3.7	−17.9
Precipitation													
Inches	0.44	0.14	0.24	0.25	0.5	0.18	0.18	0.43	0.5	0.3	0.25	0.26	3.67
Centimeters	1.1	0.35	0.6	0.63	1.25	0.45	0.45	1.08	1.25	0.75	0.63	0.65	9.18

changes might thwart our advancing technology. This could leave mankind helpless to do more than struggle weakly to keep from falling behind in the battle for survival.

Emmanuel LeRoy Ladurie, a Frenchman and an historian of agricultural problems, has pointed out the many changes in world climate and weather that have significantly affected human existence during the past 1000 years.* There have been a number of times when entire populations have been forced to leave formerly hospitable areas because of localized alterations in the weather. He cites, for example, the case of Europeans who were forced to move from Scandinavia because the weather during the eleventh century turned too cold. Another example occurred near the end of the thirteenth century, when Indians of what is now Arizona and New Mexico were decimated by a drought that lasted for 60 years. Fluctuations in climate have been an integral part of the earth's solar budget about which, until, possibly, the present time, humans could do little. Ladurie makes a strong point in favor of a relatively new approach to the study of agricultural problems, a scientific discipline called **agricultural meteorology.**

An agricultural meteorologist (or agricultural climatologist) studies the important food crops to discover which climate regions are most suitable for them. If cereal grains are under investigation, the scientist will determine what temperatures (day, night, and seasonal) are best for growing bountiful crops of wheat, corn, rye, or barley. Then he will search the world over for the places that fit the requirements. When scientists from all countries of the world cooperate in this endeavor, it may be possible to provide food at more than mere subsistence levels for all people.

One final note: It would be beneficial to future generations if the people of today were to actively pursue solutions to the problems of vegetation and climate, while at the same time applying brakes to the expanding world population.

Review and Discussion

1. Explain Wladimir Köppen's classification system and how he arrived at this method.

2. Explain the basic differences between Köppen's five major categories: A, B, C, D, E.

3. Using Table 8–1, explain the following climatic alphabetical codes: Aw, Csa, ET, BSk, Dfc, Cwa, BWh.

4. In which climate region is the northern limit of forest growth?

5. Why are highland climates difficult to classify or examine in detail?

6. Which climate regions are the most preferred for human settlement? Why?

7. What air masses dominate (a) tropical savannas, (b) continental subarctic, (c) midlatitude steppe and prairie?

8. Explain the differences between the climates of Dallas, Texas, Los Angeles, California, and Montgomery, Alabama.

*Emmanuel LeRoy Ladurie (1971), *Times of Feast, Times of Famine: A History of Climate Since the Year 1000,* trans. Barbara Bray (Garden City, N.Y.: Doubleday.

9. How do humid subtropical regions differ from humid continental regions?

10. Explain why the geographical positions of the Mediterranean dry subtropical and the humid subtropical climates cause them to be different.

11. What is the main difference between tropical dry and midlatitude dry climates?

12. Explain why cold or cool ocean currents have such an effect on climates.

13. Locate on a map the most famous savanna regions of the earth. Do the same for steppe and prairie regions.

14. What are the reasons for the different vegetation forms of the tundra and the taiga?

15. The Great Plains of North America lie in what climate region?

Note

The extremes in temperatures in various climate regions cited in this chapter were taken from the following interesting and informative article: Glenn Cunningham and James Vernon (December 1968), "Some Extremes of Weather and Climate," *The Journal of Geography,* 67, no. 9, 530–535.

Additional Reading

In addition to the citations for chapters 4 through 7, consult the following:

Appleman, Philip (1965). *The Silent Explosion.* Boston, Mass.: Beacon.

Broek, Jan O. M., and John W. Webb (1973). "Problems of Population Growth," in *A Geography of Mankind,* chap. 20. New York: McGraw-Hill.

Burton, Ian, and Robert W. Kates (January 1964). "Slaying the Malthusian Dragon," *Economic Geography,* 40, no. 1.

Court, Arnold (1973). "The Climate of the Conterminous United States," in R. A. Bryson and F. K. Hare, *Climates of North America,* chap. 3. Vol. 11 of *World Survey of Climatology.* Amsterdam: Elsevier Scientific Publishing.

De Blij, Harm J. (1974). "Infinite Humanity: The End of Evolution?" in *Man Shapes the Earth,* chap. 3. Santa Barbara, Calif.: Hamilton.

Higbee, Edward (1960). *The Squeeze: Cities without Space.* New York: William Morrow.

Johnson, Cecil, ed. (1970). *Eco-Crisis.* New York: Wiley.

Johnson, Warren A., and John Hardesty (1971). *Economic Growth vs. the Environment.* Belmont, Calif.: Wadsworth.

Manners, Ian R., and Marvin Mikesell, eds. (1974). *Perspective on Environment.* Publication no. 13. Washington, D.C.: Association of American Geographers. See especially, Werner H. Terjung, "Climatic Modification," chap. 5.

Thornthwaite, C. W. (1943). "Problems in the Classification of Climates," *The Geographical Review,* 33, no. 2, 233–255.

Trewartha, Glenn T. (1962). *The Earth's Problem Climates.* Madison: University of Wisconsin Press.

Wents, Frits W. (June 1957). "Climate and Agriculture," *Scientific American* (offprint no. 852).

——— (1963). *The Plants.* New York: Time-Life Books. See especially "The Moulding Forces of Climate," chap. 6.

9 Composition of the Earth

The whole valley, sparkling in the late sunlight, looks like a trim, polished, perfect existence. The dome Tissiack looks down the valley like the most living being of all the rocks and mountains; one would fancy that there were brains in that lofty brow. How grandly comes the gloaming over this pearly beauty!

JOHN MUIR

In 1869, John Muir, noted mountaineer and naturalist, spent his first winter in Yosemite Valley and was so overwhelmed by its spectacular beauty that he fell under the spell of the mountains and wilderness regions of the West. That he attained a feeling of oneness with the earth is evident in the words he wrote in 1890: "There is a love of wild Nature in everybody, an ancient motherlove ever showing itself whether recognized or no, and however covered by cares and duties."[1]

But Muir was not content with mere exhortations; he was a man of action, and he fought nobly to preserve our nation's forests and the rugged places he held to be almost sacred. He founded the Sierra Club and was instrumental in persuading President Theodore Roosevelt to set aside for posterity the lands that now contain Sequoia and Yosemite national parks. He tried to convince all who would listen to him that conservation could not wait. Unfortunately, many would not heed his words, impelling him to write: "Most people are on the world, not in it—have no conscious sympathy or relationship to anything about them—undiffused, separate, and rigidly alone like marbles of polished stone, touching, but separate."[2]

For John Muir, ecology meant the preservation of the earth and the ecosystems interrelated with it, but he also believed that one could not function properly as an ecologist without reaching an understanding of the earth's structure and composition.

With this idea in mind, we shall consider the earth's origin and the reasons it has evolved to its present state. We shall examine the earth's interior and exterior and learn about its basic mineral and rock composition. Then we shall consider the powerful forces that have elevated entire continents and great mountain ranges, as well as studying other forces that have operated to bring them down again. The earth's uneven and beautifully sculptured face did not get that way overnight. It took time—an almost inconceivable amount of time.

Geologic Time

Since the beginning of our existence on the earth, there have been almost as many theories concerning earth's history as there were people. Because humans witnessed the alternating periods of day and night, the phases of the moon, and the cycle of the seasons, they thought only of days, weeks, months, years, and, sometimes, in terms of millennia (thousands of years). While we of today may say that we truly understand the length of earth's history, it is difficult to comprehend the vastness of cosmic time. Can you honestly say you can conceive of 1 billion years of elapsed time?

Geologists and geomorphologists have tried to fathom the riddle of the ages by measuring the thicknesses of different layers of rocks, the salinity of the seas, or the lowering of landmass surfaces. As you might guess, however, their conclusions do not always agree. A step forward came with the discovery by England's Lord Rutherford, in 1904, that mineral elements disintegrated through time, each at a fixed rate of radioactive decay.[3] This meant that almost every substance lost radioactivity, and the determination of the various rates of decay of minerals, rocks, fossils, trees, and other bits of matter supplied geologists with a relatively accurate time clock. While geologists admit that radioactive dating is not absolutely accurate, they do claim it is close enough to serve us until a better method is discovered.

The age of the earth is now estimated to be about 4.5 billion years. Because the earth did not come into existence exactly as it is now, geologists and geomorphologists have put their research findings on a chart that shows the evolutionary stages of its development. This is termed the **geologic time scale** (Fig. 9–1). There are five major divisions termed **eras** and a number of subdivisions, called **periods** and **epochs**, each with its own special characteristics and features. The oldest era is called the Archeozoic era, and during this span of some 2 billion years, the earth's crust formed, moisture was released, and the atmosphere came into being. Then followed the Proterozoic era, which lasted for about 2 billion years, during which time the oceans appeared and the first life forms evolved in the shallow waters of a continent (called Pangaea) surrounded by the seas. These two eras are often combined and referred to as the Precambrian era.

To many people, the most exciting events of earth's history occurred after the end of the Precambrian era, about 600 million years ago, because the life forms advanced from the single-celled algae and fungi to forms we can visualize and understand: clams, fish, reptiles, grasses, shrubs, and trees. Physical events and the appearance of some dramatic new life form are the dividing factors, and the place where the event occurred or the fossil was found gave its name to the interval of geologic time. An example would be the Carboniferous period of the Paleozoic era. Forests covered vast areas of the earth through much of this period of time, and these forests have provided mankind with coal and other carbon products. Because these coal deposits were first studied in Pennsylvania and Mississippi, geologists in the United States gave the epochs those names.

Why is geologic time important? Because it gives us a better understanding of the tremendous expanses of time that must occur before continents can be raised, mountains built or leveled, valleys carved, or nearly level plains created. As we move ahead with our investigation, refer occasionally to Fig. 9–1 so you will be able to maintain the proper perspective.

Earth's Interior

You may have heard someone refer to the earth as terra firma (*terra,* "earth," and *firma,* "solid"), but is it really so solid? No, it is not. The earth's crust actually is in almost constant movement—as amply proved by the occurrence of many earth-

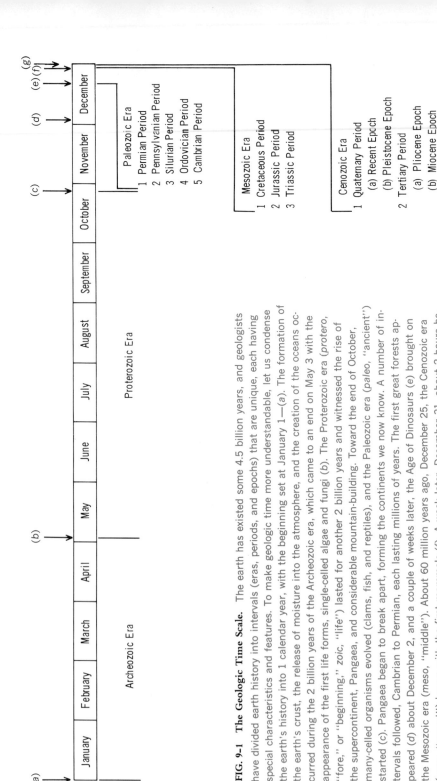

FIG. 9-1 The Geologic Time Scale. The earth has existed some 4.5 billion years, and geologists have divided earth history into intervals (eras, periods, and epochs) that are unique, each having special characteristics and features. To make geologic time more understandable, let us condense the earth's history into 1 calendar year, with the beginning set at January 1—(a). The formation of the earth's crust, the release of moisture into the atmosphere, and the creation of the oceans occurred during the 2 billion years of the Archeozoic era, which came to an end on May 3 with the appearance of the first life forms, single-celled algae and fungi (b). The Proterozoic era (protero, "fore," or "beginning," zoic, "life") lasted for another 2 billion years and witnessed the rise of the supercontinent, Pangaea, and considerable mountain-building. Toward the end of October, many-celled organisms evolved (clams, fish, and reptiles), and the Paleozoic era (paleo, "ancient") started (c). Pangaea began to break apart, forming the continents we now know. A number of intervals followed, Cambrian to Permian, each lasting millions of years. The first great forests appeared (d) about December 2, and a couple of weeks later, the Age of Dinosaurs (e) brought on the Mesozoic era (meso, "middle"). About 60 million years ago, December 25, the Cenozoic era (ceno, "recent") began with the first mammals (f). A week later, December 31, about 2 hours before midnight, mankind appeared (g). Many of the greatest achievements of mankind have occurred during 1 second of geologic time (200 years). Imagine: In 1 second of time we have moved from the steam engine to space flight. What if we had an entire year?

quakes and volcanic eruptions each year. Why must the earth experience these incessant motions? For the answer, we must look below the surface and examine the earth's deep interior (Fig. 9–2), the mantle and the core.

The earth's diameter is estimated to be about 7940 miles (12,704 kilometers), and most of this dense mass exists in an intensely heated state. At the center is the **inner core**, a solid mass of nickle and iron about 1560 miles (2496 kilometers) in diameter with a temperature of around 7000°F (4000°C). Wrapped around the inner core is a liquid layer of the same metals called the **outer core**, which has a temperature close to 6332°F (3500°C). It is from this core area that heat is constantly moving outward in what might be termed convectional streams, thus maintaining the temperature of the next region, the **mantle**, at a range of from 6332°F (3500°C) down to 3632°F (2000°C). The currents circulate throughout the mantle, tending to keep the earth and its crust in an unstable condition. Should there be an abrupt increase in the temperature of a portion of the mantle, the heated region might expand, and in some cases, it might result in a lifting of the crust above. In other cases, the upper mantle might liquefy (the rocks would melt) and volcanic activity would ensue. Sometimes portions of the earth's crust will settle where mantle rock has been displaced by the heat currents.

The upper reaches of the mantle are more pliable than the deeper regions, more responsive to the heat-circulation system within the earth. This upper zone is termed

FIG. 9–2   Earth's Interior.

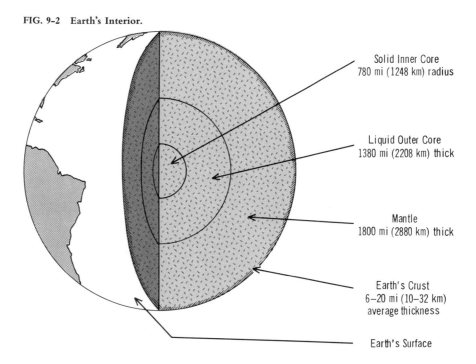

Solid Inner Core
780 mi (1248 km) radius

Liquid Outer Core
1380 mi (2208 km) thick

Mantle
1800 mi (2880 km) thick

Earth's Crust
6–20 mi (10–32 km)
average thickness

Earth's Surface

the **asthenosphere** (*asthenes,* "weak") and some earth scientists describe it as a flexible, almost plastic, layer over which portions of the earth's crust can slide (as you will see when we discuss the theory of continental drift in the next chapter). The boundary between the asthenosphere and the crust is referred to as a **discontinuity** (distinct separation) and is called the **Moho.** This term is an abbreviation of the more correct term *Mohorovičić discontinuity* (named after the Yugoslavian seismologist, Andrija Mohorovičić, who discovered it). The Moho varies in depth below the earth's surface, being much closer to the earth's surface beneath the oceans than under the continents (Fig. 9–3). It is an uneven boundary, reflecting the weight of the masses above it.

As yet, no one has penetrated the Moho, although the scientific world is understandably enthusiastic about such a project; they want to know more about the mysterious mantle.* It is believed that most of the earth's internal movements begin within this region, which is some 37 to 156 miles (60 to 250 kilometers) below the surface.

Earth's Crust

The ancient Greeks had the idea that the earth was composed of four elements: fire, water, air, and earth. When we think of the earth, which is our home, we think of four elements, too, but not the same ones. We call the air our atmosphere; the earth's waters, our hydrosphere; the life zone, our biosphere; and the earth mass itself, the lithosphere.

The term **lithosphere** is derived from the Greek *lithos,* "rocks," and *sphaeros,* "sphere," thus, a "rocky sphere." Some consider the lithosphere to include the entire earth, but most earth scientists use the term to define and delimit the region above the Moho. This is a region of less pliable rock materials that comprises two distinct layers. The bottom layer rests directly on top of the asthenosphere and underlies the continental landmasses. It has been given a name based on its most common minerals, silicon and magnesium, thus **sima,** and it is composed of heavy, dense basaltic rocks. The upper layer is made up of lighter and less dense granitic rocks, and its most common minerals, silicon and aluminum, have earned it the name of **sial,** or the **sialic layer.**

The earth's crust varies considerably in thickness. Where the crust includes only the sima, as under the world's oceans, it is only about 5 miles (7.5 kilometers) thick, but where the sial and the sima are together, the thickness increases to about 20 miles (32 kilometers) as an average, and as much as 30 miles (48 kilometers) under great mountain ranges. The highest mountain, however, has an elevation amounting

---

* For an interesting explanation of the earth's interior and its rather plastic mantle, see: Don L. Anderson (July 1962), "The Plastic Layer of the Earth's Mantle," *Scientific American,* 209, 52–59. Also available as offprint no. 855.

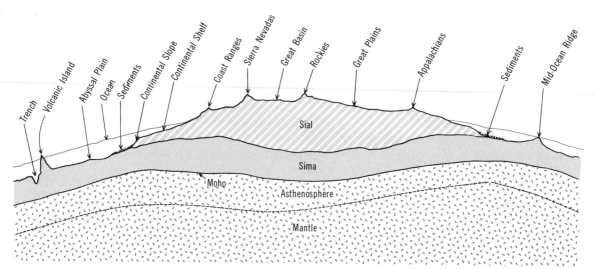

FIG. 9-3 The Earth's Outer Margin: Crust and Upper Mantle. This is an exaggerated (not-to-scale) cross section of the United States at approximately 38° north latitude.

to about half the difference between the two averages. Thus, Mt. Everest measures slightly over 5 miles (8 kilometers) in height. Both the sial and the sima are affected by heat currents operating within the mantle and by other forces operating there. We shall continue this discussion of earth movements in chapter 10.

Earth Materials

Many times during our discussion of the composition of the earth we have mentioned rocks. We have talked about mantle rock, the rocks of the earth's crust, heavy and dense basaltic rocks, and lighter, less dense granitic rocks. But, what is a rock? Perhaps you can name some, such as granite, slate, limestone, and sandstone, but do you know of what they are made? How did they come into being? How many kinds of rocks are there? To arrive at some answers to these questions, we must start at the beginning. Rocks are composed of minerals. Some rocks have only one mineral; some have combinations of two or more.

MINERALS **Minerals** come in a variety of sizes and chemical combinations and have a variety of physical properties. **Mineralogists** (scientists who study minerals) have identified over 2000 different minerals, combinations of more than 96 elements that exist in and on the earth. A mineral can be made up of one element (such as gold or silver) or a combination of two or more (such as quartz, which contains silicon and oxygen). And what is an element? An **element** is matter that cannot be broken down or further subdivided by natural means. Of the many elements that are present in the earth's crust, the most abundant (Table 9–1) are silicon (28 percent) and oxygen (47 percent). This does not mean that the other elements are not significant. Some are scarce and cannot easily be discovered and mined, but all are important, because they make up most of the weight of the earth's crust.

Elements are made up of atoms, and each atom in a particular element shares in the physical and chemical characteristics of that element. We could go further into the mysteries of the atom and learn how it is made up of neutrons, protons, and electrons, but we shall not. We must leave that to geophysicists and atom smashers. We can, however, tie it all up in a kind of equation:

Atoms form elements.
Elements form minerals.
Minerals form rocks.

Minerals can form in a number of ways: by crystallization from evaporation (when the moisture evaporates, leaving behind a mineral such as salt from sea water), by crystallization from a molten state (when once-fiery liquid rock, called **magma,** cools and allows minerals such as quartz or feldspar to form), and by condensation from a vapor (when, for example, sulfur is heated to a gaseous state by the intense heat within the vent of a volcano and, as it escapes, cools and condenses into beautiful yellow crystals around the mouth of the vent).

While some rocks can be made up of only one mineral (such as gold), most rocks are a combination of two or more. Quartz, feldspar, and mica are the minerals that make up granite, and if you pulverized a chunk of granite, the result would be tiny pieces of quartz, feldspar, and mica. Why? Because each mineral always retains its own, separate characteristics. When it is released from combination, it resumes its own identity.

TABLE 9–1 The most abundant mineral-forming elements (percentages rounded off) in the lithosphere

ELEMENT	CHEMICAL SYMBOL	PERCENTAGE (BY WEIGHT)
Oxygen	O	47 ⎫ 75
Silicon[a]	Si	28 ⎭
Aluminum	Al	8.1
Iron	Fe	5
Calcium	Ca	3.6
Sodium	Na	2.9
Potassium	K	2.6
Magnesium	Mg	2.1

[a] The most abundant mineral in the lithosphere is quartz, which is composed of silicon and oxygen. If nothing is added, quartz will look like pure, clear glass.

Mineralogists have tried to simplify matters by considering the thousands of mineral combinations as "families." They group together all mineral combinations that have the same or similar basic composition (Table 9–2). But, is that enough? Some geomorphologists want to have a more thorough understanding of the part of earth they are studying, so they want to know the minerals contained in the rocks of the area. They examine the physical properties of the rocks by performing one or a number of tests. They may test for hardness, to discover what it takes to scratch a mineral. Talc is the softest of all the minerals and can be scratched with a fingernail. Diamonds are the most difficult to scratch, and in between (in ascending order of hardness) are gypsum, calcite, fluorite, feldspar, quartz, topaz, corundum. You can also test by examing the mineral's luster (whether it is shiny like glass or is pearly, metallic, silky, fibrous, or dull), its cleavage (how it breaks up when hit by a hammer, whether it shatters or splits into sheets or fibers), or by visual inspection (how it looks to the naked eye).*

ROCKS   Rocks, of course, come in a variety of sizes, shapes, and colors.† There are solid-appearing granite cliffs, accumulations of rock debris along a highway cut through a mountain or hill, rounded pebbles at the bottom of a stream, and black sand on a beach in Hawaii. **Petrologists** (earth scientists who specialize in the study of rocks) have neatly classified the rocks of the earth into three classes: igneous, metamorphic, and sedimentary.

Igneous rocks   **Igneous** rocks are the result of volcanic activity. They are rocks "born of fire," rocks that were once in a molten state and hardened as they cooled. If

---

*For more detailed information on minerals and how to recognize them, see: A. M. Bateman (1950), *Economic Mineral Deposits* (New York: John Wiley); W. Ernst (1969), *Earth Materials* (Englewood Cliffs, N.J.: Prentice-Hall); and B. Simpson (1966), *Rocks and Minerals* (New York: Pergamon Press).

†To be considered a true rock, an object does not have to be a solid chunk of granite, a pebble, or a boulder. A handful of sand is also a rock. If it is a solid mass, it is called *consolidated;* if it is loose material, it is termed *unconsolidated*.

TABLE 9-2   The basic mineral combinations

FAMILY	COMPOSITION
Oxides	Among oxides, oxygen is the prime element, and it is usually united with another element. As an example, quartz is composed of silicon and oxygen.
Carbonates	Carbonates occur when oxygen and carbon are united. If calcuim is added, calcium carbonate results, one form of which is limestone.
Sulfides	Sulfides are a combination of sulfur and oxygen. If another mineral, such as calcium, is added, a combination called gypsum would be the result.
Silicates	Silicon is the primary element in silicates. If oxygen is added, quartz results. If iron is added, along with magnesium, hornblende results.
Chlorides	Chlorine is the dominant element in chlorides. If sodium is added, sodium chloride (common table salt) results.

(a)

(b)

(c)

(d)

Intrusive (Plutonic) Igneous Rocks. (a) Devil's Marbles, round granite boulders in Australia's Northern Territory; (b) granite, an intrusive (plutonic) igneous rock; (c) columns of gabbro in the Hudson Palisades, New York; (d) gabbro, also called diabase, an intrusive (plutonic) rock.

molten rock (magma) works its way upward from the asthenosphere into the earth's crust and finds its passage blocked, the magma must cool deep within the crust, and large-grained crystals will have time to form. Granite, gabbro, and diorite are examples of what is termed **intrusive** igneous rock. For igneous rocks that form underground, some petrologists prefer the term **plutonic** (from the ancient Greek god of the underworld, Pluto).

When magma escapes onto the outer surface of the earth in the form of a lava flow or in a violent blast of ashes or blobs of lava, the material must cool rapidly, and there will be insufficient time for large crystals to form. This type of igneous rock is termed **extrusive.**

While color does have some value in identifying different kinds of igneous rocks, the grain or crystal structure is the most important difference. For example, the most common intrusive igneous rock is granite, which is large grained; the most common intrusive rock is basalt, which is fine grained. The Hawaiians have given interesting names to the lava flows of basalt on their islands. If the moving, or once-moving, mass is rough textured and "blocky"—a jumbled pile of black rocks—they call it *aa* (pronounced *ah-ah*). If the flow is rather smooth-looking and ropelike, they call it *pahoehoe* (pronounced *pah-hoey-hoey*). Ashes blown from a volcano do not look like a rock until they have had time to solidify and compact into a rock called *tuff*. Italy is a land of volcanic activity, and the ancient Romans constructed many buildings in Rome with bricks carved out of compacted tuff.[4]

Sometimes blobs of magma are spouted from a volcano's crater. These red-hot gobs of lava look like hand grenades or bombs, so they are called *volcanic bombs*. Sometimes lava splashes out of the crater, showering the vicinity with splotches of magma that cool so rapidly that there is no time for crystals to form. The result is a textureless black rock that looks like black glass. This is obsidian, called *volcanic glass*. Extrusive rocks that have been blown from a vent of a volcano are called **pyroclastic** ("broken up by fire") and **ejecta** (having been "ejected" from the volcano). Another kind of ejecta is a light grayish frothy-looking rock that is so filled with air spaces that it can float in water. This is pumice.

Sedimentary rocks **Sedimentary** rocks result from sediments (sand, silt, clay, or the skeletal remains of once-living organisms) deposited by wind, stream, lake, ocean, or glacier. The sediments are deposited grain-by-grain on an open plain, a valley floor, a stream or lake bottom, or an ocean floor. After the grains have been chemically cemented together or compacted by the pressure of other layers above, they become known as sedimentary rocks. So common are these rocks that they make up about 75 percent of all the rocks on the earth's surface. They can be further classified as clastic, organic, and chemical.

Clastic sedimentary rocks are made up of rock fragments that range in size from boulders down through cobbles, pebbles, sand, silt, and clay. The type of clastic rock called *sorted* refers to *shales,* which are compacted flakes of clay or silt particles that easily split into thin, fine-grained layers, and *sandstone,* which is composed of grains of sand cemented together. Shales are the most abundant sedimentary rock, with sandstone second. The type known as unsorted refers to compacted mixtures of various sized particles, with small boulders, pebbles, and sand or silt combined to form a *conglomerate.*

Organic sedimentary rocks are derived from the skeletal remains of once-living creatures. Examples include calcareous limestone, made up of shells and coral; coquina, made from shell fragments; reeflimestone, made from the excretions and bodies of coral; and chalk, made from microorganisms that are invisible without magnification. Coal is another form of organic rock, derived from once-living trees and other plant forms that carbonized through stages: from bogs to peat to coal.

(a)

(b)

(c)

(d)

(e)

(f)

Extrusive Igneous Rocks. (a) is a photo of an aa lava flow north of Hatch, Utah. (b) is a pahoehoe lava flow in the Craters of the Moon National Monument, Idaho. (c) shows deep deposits of tuff in Big Bend National Park, Texas. Note the ejecta and volcanic bombs in the foreground. (d) is a close-up of obsidian, pyroclastic ejecta also known as volcanic glass. Pumice from Sunset Crater, Arizona, is shown in (e). (f) is basalt.

(a)

(b)

(c)

(d)

Sedimentary Rocks. (*a*) "Sandstone Symphonies"—cross-bedding patterns of the sedimentary rock called sandstone, in Zion National Park, Utah; (*b*) conglomerate rock in a road cut near Suffern, New York; (*c*) petrified wood protruding from beds of sandstone in Arizona's Petrified Forest; (*d*) coquina, a sedimentary rock made up of seashells; this outcrop is in the Fezzan Desert in Libya.

Chemical sedimentary rocks are the result of precipitation of dissolved chemicals through layers of rock or from water in which the chemicals were suspended. For example, table salt (sodium chloride) results from the evaporation of water from a salty, liquid solution. After the water is gone, the salt remains. Another chemical rock that results from evaporation is gypsum, a common building material used in plaster or in drywall construction. Salt and gypsum are often called **evaporites.** While limestone can result from the skeletons of once-living creatures, it can also result from the addition of carbon dioxide, carried to the earth by rain. If carbon dioxide combines with calcium, calcite (calcium carbonate) is formed. When underground water carries the calcite in solution down to the roof of a cavern, stalactites and stalagmites (see chapter 13) are formed. This type of chemical rock is called *travertine*. Another interesting form of chemical rock is *petrified wood,* which results when chemicals replace the original woody cells of a fallen tree.

Sedimentary rocks are especially interesting to geologists, geographers, archaeologists, as well as to rockhound hobbyists, because only in these rocks do we find fossils and other features that give us insight into the history of our earth. The layers of rocks, called *beds,* or **strata,** can be read almost as though they were pages in a book by those who understand the "language." In some of these beds we find *ripple marks* (tiny "wavelets" that were pressed onto the surface of a sand dune or the floor of a lake or ocean, then covered up by new deposits), footprints of dinosaurs or birds, *geodes* (hollow rocks usually occurring in limestone formations and having beautiful crystals jutting from the inner shell), and, of course, fossils of ancient creatures or plants.

Metamorphic rocks **Metamorphic** rocks were either sedimentary or igneous when formed, then were either buried very deep within the earth's crust where they were later subjected to great pressures, or they were subjected to extremely high tempera-

Fossils of a Long-Ago Age.

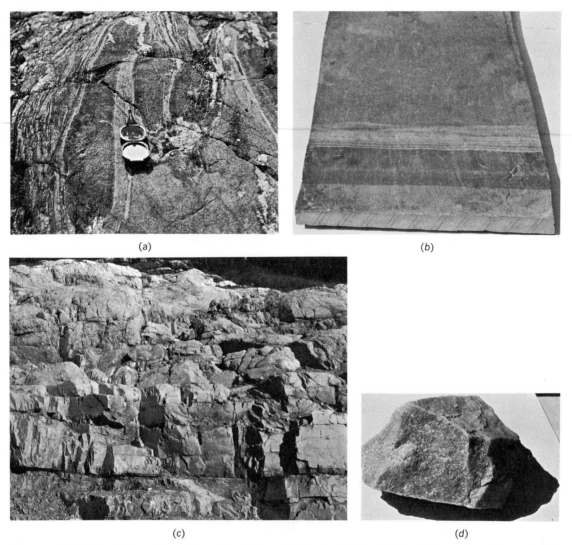

(a)

(b)

(c)

(d)

Metamorphic Rocks. (a) banded metamorphic rock along the Hudson Highlands, New York: granite metamorphosed into gneiss and twisted into bands by the intrusions of a rock called pegmatite; (b) slate from eastern Pennsylvania: a popular flooring material that results from the metamorphosis of shale; (c) fractured and jointed blocks of marble (metamorphosed limestone) in a road cut near South Dorset, Vermont; (d) quartzite, metamorphosed sandstone.

tures from a nearby mass of magma. As a result, changes were produced, and the rocks underwent metamorphosis (from the Greek *meta,* "changed," and *morphe,* "shape" or "form"). The original texture and minerals are altered, and the new rock has different physical and chemical characteristics. One example of a metamorphic rock is the industrial diamond, which was metamorphosed by heat and pressure from coal into pure carbon.

The transforming influence of heat is called **contact metamorphism,** which means, simply, that an igneous body of rock supplied sufficient heat to bring about changes in rocks nearby. Granite can become altered into a foliated (the appearance of plates or thin leaves) gneiss. Then there is metamorphism brought about by pressure, **dynamic metamorphism.** Great pressures constantly are being exerted on rock masses throughout the earth's crust, along vast breaks or fracture planes and through a variety of crustal movements. Here, we do not find radical chemical changes, but we do discover new forms, new textures, and a realignment of the existing minerals into new crystal structures.

There are, of course, some alterations produced by the movement of chemically active solutions that dissolve existing chemicals, add new elements, and thus introduce a new structure with new crystallizations and shapes. For example, a rock called gabbro is transformed into another called serpentine by chemically active waters that have been superheated by intrusions of nearby magma.

THE ROCK CYCLE   As noted earlier, there are a number of cycles that have an important bearing on planet earth. We have discussed the cycle of the seasons, the solar energy and heat balance, the hydrologic cycle, and the cycle of vegetation, and now we shall study the **rock cycle** (Fig. 9–4).

FIG. 9–4   **The Rock Cycle.**

The rock cycle began billions of years ago, as the earth cooled and the crust began to form, and it has endured throughout earth's history. When moisture was released by the cooling earth and the atmosphere allowed rain to fall, alternating periods of excessive and drying heat and heavy precipitation caused the newly forming igneous rocks to be weathered (a process we shall consider further in chapter 11) and broken down. Streams carried the smaller pieces down from the emerging elevations to lakes, oceans, and valley floors. Ashes, spewed forth by volcanic activity, settled on the land and in the waters. As time passed, these sediments were built up into layers, and the sediments were compacted and cemented together by chemicals in the infiltrating waters. Sedimentary rocks were formed, and as more time passed, the tremendous weight of overlying beds created pressures that worked, along with the heat penetrating upward from the mantle, to change the mineral structures of the rocks. They became transformed into metamorphic rocks. Earth movements twisted the beds and folded them, and once again heat from the earth's interior caused the rocks to melt. Igneous rocks were created out of the old.

The cycle went on through the ages, and occasionally there were interruptions in the cycle, as sedimentary rocks bypassed the perpetual round and were melted into magma without going through the stage of metamorphosis. Igneous rocks were subjected to intense heat and pressure and were transformed directly into metamorphic rocks. Metamorphic rocks were attacked by weathering agents and broken into fragments that eventually ended up as sedimentary rocks. There appears to be no end to this cycle. It still goes on.

Review and Discussion

1. Describe the composition of the atmosphere, hydrosphere, biosphere, and lithosphere.

2. Make a sketch of the earth's interior, and label the major zones. What are the distances involved?

3. What does the term *Moho* mean?

4. How did the geologic eras, periods, and epochs get their names?

5. Discuss these terms: *Archeozoic, Proterozoic, Paleozoic*. What types of life forms existed in each?

6. What is the mantle, and how does it differ from the earth's core?

7. What is the asthenosphere?

8. What is the difference between a mineral and a rock?

9. What are elements?

10. Which is the most abundant mineral?

11. Describe some of the different ways minerals are formed.

12. Discuss these mineral combinations: oxides, carbonates, sulfides, silicates, and chlorides.

13. What are some of the tests used to understand the different physical properties and structures of minerals?

14. What is magma? How does it differ from lava?

15. What is the difference between aa and pahoehoe lava?

16. Define these terms: *pyroclastic, ejecta, volcanic glass.*

17. How many kinds of sedimentary rocks are there? Explain the differences.

18. What is the most abundant sedimentary rock?

19. What is the difference between the terms *organic* and *inorganic?*

20. What kind of rock is common table salt?

21. What is the difference between contact and dynamic metamorphism?

22. What terms are used to describe the layers of rocks?

23. What kind of rock is an industrial diamond?

24. When limestone is metamorphosed it becomes what kind of rock?

25. Discuss the rock cycle.

Notes

[1] John Muir (1938), *John of the Mountains: The Unpublished Journals of John Muir,* ed. Linnie Marsh Wolfe (Boston, Mass.: Houghton Mifflin), p. 315. With permission.

[2] Muir, p. 320. With permission.

[3] William C. Putnam (1964), *Geology* (New York: Oxford University Press), p. 431.

[4] Putnam, p. 52.

Additional Reading

American Geological Institute (1962). *Dictionary of Geological Terms.* Garden City, N.Y.: Doubleday.
Beiser, Arthur (1962). *The Earth.* New York: Time-Life Books.
Deer, William A. (1966). *Introduction to the Rock Forming Minerals.* New York: Wiley.
Ernst, W. (1969). *Earth Materials.* Englewood Cliffs, N.J.: Prentice-Hall.
Longwell, C. R., R. F. Flint, and J. E. Sanders (1969). *Physical Geology.* New York: Wiley.
McKenzie, G. D., and Russell O. Utgard, eds. (1972). *Man and His Physical Environment.* Minneapolis, Minn.: Burgess.
Shelton, John S. (1967). *Geology Illustrated.* San Francisco: Freeman.

10 Land Surface and Sea Bottom

When we dwell with mountains, see them face to face, every day, they seem as creatures with a sort of life—friends subject to moods, now talking, now taciturn, with whom we converse as man to man. They wear many spiritual robes, at times an aureole, something like the glory the old painters put around the heads of saints. Especially is this seen on lone mountains, like Shasta, or on great domes standing single and apart.

JOHN MUIR

We shall now center our attention on the earth's surface features, the landforms that rose out of the mantle and were twisted, distorted, fractured, elevated and lowered, and shifted horizontally. We must keep in mind that mountains, valleys, hills, plateaus, and plains are not limited to the continents. Features present on the continents are also present on the ocean floors. If the waters of the oceans were to be drained away, the resulting landscape would appear much like that of the landmasses. Why is this? Why are there deep valleys, gaunt fractures, massive mountain ranges, and broad plains on the sea bottom? Are they caused by waves and currents? The answers to these questions can only be answered by examining the whole—the land surface and the sea bottom.

The same forces that moved the continents, elevated mountains, caused volcanoes to spew forth their fiery outpourings, and fractured the land are operating throughout the earth's crust, and they have done so since the beginning. Later in this chapter you will see how much alike the land surface and sea bottom really are, but for now, let us concentrate on the internal, or tectonic, forces involved in the shaping of the earth's surface.

Tectonic Forces

Today, people tend to be scientific in their approach to the unknown. When the earth shivers and shakes, as it often does in Italy, Peru, Central Asia, or Nicaragua, or when volcanoes erupt, as they do in Mexico, Hawaii, or off the coast of Japan, earth scientists speak of diastrophic forces, vulcanism, and plate tectonics. People of earlier times, however, thought that a god or earth spirit was showing his strength or his displeasure at their behavior. The Japanese once called the "earth shaker" Nai-no-Kami (god of the restless earth) and built temples in his honor—hoping that by so doing they could placate his wrath.[1] The people of Nicaragua worshipped a goddess of volcanoes, Masaya, and threw human sacrifices into a crater to appease her.[2] The volcanic island of Lemnos in the Mediterranean gave rise to the cult of Haphaestus, a thunder god, god of the fires of the earth. He appears in Greek mythology as a beneficent god tending the earthly fires that allowed men to work with iron and other metals.[3] The Romans of later times adopted this god and called him Vulcan. He became their god of volcanoes and the home fires, and his name is now associated with the science of vulcanism, which studies this phenomenon.[4]

Such simplified answers to the question of the earth's restlessness may have been acceptable to those people of earlier times, but they have not satisfied earth scientists for several centuries. Many theories have been conceived by geophysicists and geologists trying to explain the tremendous forces and processes. While some disturbances and distortions within the earth's crust are easy to understand, others are not. Cause and effect are obvious when a section of earth collapses over a mine shaft, but how can we conceptualize the unleashing of power immense enough to lift a great mountain range or to split an ocean basin?

CONTRACTION   The most widely held theory of the nineteenth and first half of the twentieth centuries attributed the crumpling of the earth's surface to **contraction,** the cooling of the outer crust as the earth changed from a molten mass to a relatively solid body. As the earth released its moisture to the atmosphere and began shrinking, the process was equated to what happens to a plum that becomes transformed into a prune as it dries.

Proponents of this theory claim that the earth's interior is still cooling and reducing in size, so the earth's crust is constantly making adjustments to the new conditions. As the pressures mount in one place, a continent can be elevated and arched upward in its center or along one side, and in another area a landmass can be compressed downward as its edges are rumpled into a series of mountains. A state of equilibrium can occasionally be attained, but it does not last, for the cooling and shrinking is an ongoing process.

CONVECTION   The theory of **convection** might also be referred to as expansion, distension, or swelling, for it suggests that the earth is enlarging because of an increase in the temperature within the interior. As the heat generated by radioactive decay of minerals in the mantle rises to the surface, the materials of the mantle must expand, thus forcing continents to lift, tilt, or fracture. Not all earth scientists accept the idea that our planet is swelling, but many do agree that there are convectional currents circulating throughout the mantle. As heat is built up within the lower mantle, differences in temperatures of the upper and lower zones cause currents to flow through the more plastic layers within and near the asthenosphere. Rock materials in the simatic layer (sima) are affected by the intense heat below and are sometimes forced to move either vertically or horizontally. If the earh's crust were not so rigid and prone to break or fracture, the main result would be some volcanic activity or warping and folding of the layers above the moving masses of the asthenosphere.

While the early proponents of the convection theory tended to limit its influence to the creation of mountains, the folding of rock strata, the uplift of landmasses, and the cause of earthquakes, the earth scientists of this century have shown that there is more to this theory than formerly suspected. This idea of constantly moving convectional currents leads us to another theory, that of plate tectonics, a theory that is rapidly gaining support by scientists all over the world.

PLATE TECTONICS   The term **tectonics** is defined as "the internal forces of the earth that cause crustal movements" and comes to us from the Greek, *tectonicus,* "builder." The term **plate** means, of course, a "thin slab or piece of some material," but in this instance plate refers to the subdivision of the earth's crust into huge sections that are separated by sometimes violent fracture (or rift) zones. There are at least 10 or 12 major plates and a number of smaller ones. The plates extend down into the asthenosphere and are from 30 miles (48 kilometers) thick under the oceans and up to 60 miles (96 kilometers) thick when measured through the landmasses. Figure 10–1 indicates the present-day positions of the most important plates, and you can see that each continent occupies its own plate. Later, when we discuss earthquakes and volcanoes, you will also see that the greatest numbers of each of these forces are found along the edges of these plates. This tells us something, does it not? There must be tremendous activity occurring in those areas, activity that has been going on for a long time. But, when did it all begin? What is the existing evidence to prove that there are great plates and not merely powerful stress areas? Do these plates move, and if so, what was their original location? The answers can be found by examining the theory of continental drift.

The implication of the **theory of continental drift** is that the continents have been, and indeed still are, moving. This concept had been considered and rejected by a number of scientists through much of the nineteenth century, but it was only after a German meteorologist, Alfred Wegener, presented his views in an address to the Geological Association in Frankfurt, Germany, in 1912 that the idea became somewhat accepted.* While some earth scientists gave grudging agreement, others waged a series of stormy debates as they clung to the older theories. Extensive research conducted on both land and sea during the period from 1930 to the late 1950s silenced much of the opposition, and the idea that there was once a single supercontinent is generally accepted today. Some refer to this immense body of land surrounded by the waters of the Proterozoic world as the mother continent, because "she" later gave birth to the six continents that now exist. The more proper name, however, is the one applied to it by Wegener: Pangaea (from the Greek *pan,* "all," and *gaea,* "earth").

Pangaea is thought to have been surrounded by the world ocean some 2.5 billion years ago (near the end of the Archeozoic era), but its original position is not known. It is thought that Pangaea shifted around, breaking apart, coming back together, changing position periodically as it was pushed by the pressures exerted on the earth's crust from below. Field researchers have supplied a considerable amount of evidence to back up this claim. For example, scientists traveling through southern Africa, the Deccan Plateau of India, and parts of Australia (regions that are far from the present-day south-polar icecap and that are not believed to have been subjected to glaciation by an ice age) have found apparent proof that these lands were once positioned near the South Pole.

Another important discovery was made in the early 1960s, when it was learned that as rocks form iron particles within them will line up like a compass needle and point toward one of the magnetic poles. Scientists now working in the field of **paleomagnetism** (study of the earth's ancient magnetic fields) found that lines of magnetism in ancient rocks of the different continents were not aligned toward the

*To read a first-hand account of the research conducted by this pioneer in geophysics, see: Alfred Wegener (1966), *The Origin of Continents and Oceans* (New York: Dover Publications). This is a translation of Wegener's original work, which was first published in 1924 and revised in 1929.

FIG. 10-1 The Earth's Plates.

276

present-day magnetic poles. This could mean that the north and south magnetic poles either had traded positions or had wandered about over vast distances. More probably, the continents changed position. By carefully measuring the magnetism of rocks from the ocean floors and the landmasses, then feeding the data into computers, it was clearly shown that, if the continents could be pushed together, the lines of magnetism would point fairly close to a common set of magnetic poles of ancient times. This does not imply that the magnetic poles do not move. Rather, it indicates that such movements are not extensive.

Physical oceanographers have long studied the ocean floor, and they could not account for the sediments on the sea bottom being less thick than they should have been considering the billions of years that materials from the landmasses had been washed or blown into the oceans. Rather than an expected 10 miles (16 kilometers) depth, the sediments average out to be about 0.5 mile (0.8 kilometer) thick. Especially designed oceanographic ships found that the floor of the oceans was composed of basaltic (igneous) rocks that formed the simatic layer of the earth's crust. In addition, the ocean floor was found to have extensive mountain ranges, high plateaus, and deep "valleys" (called **deeps,** or **trenches**). Enough proof was finally in hand to convince all but the most skeptical that our continents were once a part of Pangaea, the supercontinent on a superplate.

Based on the evidence, we can now say that Pangaea began its final breakup some 180 to 200 million years ago (Fig. 10-2). The first separation resulted when the northern section (Laurasia) broke free along a line that ran through the Mediterranean into what is now the Indian Ocean. The southern portion (Gondwana) began to split up, and one section, consisting of Antarctica, India, and Australia, swung away to the southeast, while South America began to drift westward. Africa appears to have moved slightly eastward, shifting somewhat in its north-south alignment. In time, Laurasia began to weaken under the irresistible pressure. North America began driving westward, and the North Atlantic Ocean came into being. Gradually, all of the continents began to move into their present positions.

It must not be assumed that a state of equilibrium has been reached, for the prevalence of volcanic activity and earthquakes is convincing proof that continental drift is continuing. One example can be seen by observing California's San Andreas fault, which coincides with the boundary line between the American and Pacific plates. This is a violent fracture zone that is 600 miles (960 kilometers) long and extends about 20 miles (32 kilometers) down into the crust. Earthquakes occur regularly along this fault line, and it is estimated that the western side of the fault is moving northwesterly at a rate of about 1 inch (2.5 centimeters) per year. This means that someday (perhaps about 28 million years from now) the city of Los Angeles will pass San Francisco as it travels north!

Epeirogeny The movement of the earth's plates created the continents, and we use the term **epeirogeny** (from the Greek *epeiros,* "mainland," and *genesis,* "origin") when we talk of continent building. Such vast crustal movements usually take the form of widespread sinking, tilting, or uplifting of either an entire continent or a portion of it. A good example can be seen along the west coast of California, where

- - - - - - - - - - Boundaries of earth's plates   ⟵ Direction of drift   Continental shelves— over 100 million years ago

FIG. 10–2 The Breakup of Pangaea.

the shoreline that had long been subjected to the work of ocean waves and currents was tilted upward so that wave-cut platforms (marine terraces) are now hundreds of feet above sea level. Later in this chapter we shall examine additional examples of this force called **diastrophism** (from the greek *diastrophe,* "distortion") that faults, folds, and warps the land.

Orogeny **Orogeny** (from the Greek *oros,* "mountains") is another result of the moving plates and refers to the process of mountain building. Figure 10–3 illustrates the uneven surface of the earth, and a close examination will reveal the mountains and plateaus created by epeirogenic movements as well as by orogeny. If you compare this map with that of Fig. 10–1, you will see that the mountains appear most often along or close to the edges of the plates.

Orogeny occurs when the lighter and less dense crustal combination of sial and sima is thrust against the heavier simatic crustal plate. It is almost like an irresistible force meeting an immovable object. Something has to give. Sometimes the moving plate will plunge downward at the boundary line, diving into the upper mantle, and the resulting friction and increased heat causes some of the sima rocks to melt and thrust upward, thus creating volcanic mountains. The Mid–Atlantic Ridge that extends from Iceland on the north down through the middle of the Atlantic Ocean some 12,000 miles (19,200 kilometers) to a point midway between the southern tips of South America and Africa contains many volcanic islands created in this manner.

California's San Andreas Fault.

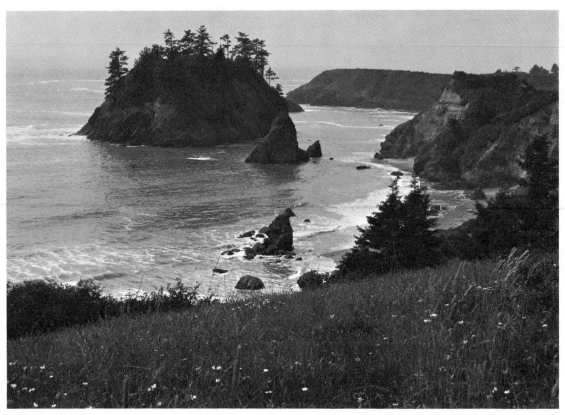

Shoreline Elevated by the Power of Epeirogeny.

Sometimes a moving plate will buckle along or near its leading edge, and the upfolding of the sialic layer will create a mountain range, such as the Andes of South America; the Rocky Mountains, Sierra Nevada, and Coastal ranges of western North America; the Atlas Mountains of North Africa; and the Himalayas of Asia. In this latter case, the Himalayas were the result of the collision of the subcontinent India with the Asian landmass. India is believed to have been a part of a segment of Pangaea that included Australia and Antarctica, but it eventually broke loose and was pushed northward by the powerful forces of the moving plates through the ancient sea of Tethys, what is now the Indian Ocean. It must have taken millions of years to accomplish this task of elevating a former sea floor to heights of more than 25,000 feet (7500 meters). Mountain climbers have reported finding sedimentary deposits containing fossils of ancient sea life on top of Mount Everest!

Isostasy   Another term often used by geomorphologists is **isostasy** (from the Greek *iso,* "equal," and *stasis,* "to stand"). It implies a tendency toward equilibrium or balance between the lighter sialic (granitic) landmasses and the heavier basaltic ocean floor. If, for example, a heavy layer of sediments is deposited through time on the

FIG. 10-3 Land Surface and Sea Bottom.

The Himalayas. The Himalaya mountain system, which has resulted from the upward force of a moving crustal plate, consists of three main ranges. Amidst one, the Greater Himalayas, on the border of Nepal and Tibet (China), stands Mount Everest, the highest mountain on earth. This photo was taken from the *Apollo 7* spacecraft.

ocean floor, that portion of the crust will tend to sink, thus forcing the nearby landmass, which has lost materials by erosion, to rise a sufficient distance to maintain the balance between the two. If a great continental glacier were to move onto a landmass, the resulting increase in mass and weight might force the sialic landmass to be depressed. The sima of the ocean floor would then have to rise in compensation. When the glacier melted and disappeared, the landmass would be relieved of its burden and probably would return to its normal elevation. The ocean floor would gradually sink until all was in balance once again.

DIASTROPHISM We have been discussing large-scale crustal movements, such as the drifting of the earth's plates, the raising of continents, and the building of mountains, all the result of the tectonic forces of diastrophism and vulcanism. Now we shall focus our attention on the smaller-scale actions of these forces and examine how they have altered the earth's structure.

Diastrophism implies crustal movement, and it can be defined as the distortion of the rock layers of the earth's crust by twisting (warping), folding (bending), and faulting (breaking) (Fig. 10–4). Irregularities created by these distortions are difficult

FIG. 10-4 Forms of Diastrophism. (on facing page) (*a*) folding: upfolds (anticlines) and downfolds (synclines); (*b*) warping; and (*c*) faulting.

(a)

(b)

(c)

Direction of dislocation or distortion   Direction of tectonic pressures to identify visually from the air or while driving across the land, unless you happen to be on a highway cut through a hill or mountain. The rupture or dislocation of rock layers is then laid out for you like a textbook diagram.

Geologists have taken core samples (by drilling down through many layers of rock) and thus are able to tell us what lies beneath our feet. The hills of Tennessee are surface features created by **folding.** This is the result of pressures exerted upon a section of land from two sides. The central portion was compressed, and the rock layers began to shift. Some found release by moving up (creating an upfold, or **anticline**); some sunk (creating a downfold, or **syncline**). If the horizontal (lateral) pressure were great enough, the layers of rock thrust upward and forward, as happened in the region of northwest Wyoming when the Grand Tetons were formed.

When the upper rock layers of the earth's crust are unable to withstand the pressures applied by the tectonic forces, one section may move past another as the rock layers fracture. This is called **faulting** (Fig. 10–5). If the movement is primarily vertical and the hanging wall portion has lowered (or the footwall has been raised), the rift or break is called a **normal** fault (Fig. 10–5[b]). If the opposite has occurred, the result is termed a **reverse** fault (Fig. 10–5[c]). If the dislocation is primarily horizontal (sideways), it is called a **strike-slip** fault (Fig. 10–5[d]). This does not mean that such crustal movements are limited to either vertical or horizontal movement. In most cases, the movement will be a combination of both. For example, the San Andreas fault is considered a "master fault," because there are a great number of smaller faults associated with it. It is popularly thought of as a massive fracture that operates in a strictly horizontal direction, from the southeast to the northwest; however, the western side, which is a part of the Pacific plate, is rising. Measurements taken after the February 1971 earthquake, resulting from movement along this fault, showed that the western side had been uplifted approximately 4 feet (120 centimeters).

Interesting landforms can be created by these crustal movements, such as hills or mountains created by vertical uplift of a central fault block (Fig. 10–5[e]) or by the tilting of one side of a crustal block. The landform created by this type of movement is called a **horst** (a German word meaning "ridge"), an example of which is the Sierra Nevadas of California. If the block between two fault systems drops because of the compression applied by tectonic forces, the result is a valley (Fig. 10–5[f]), called a **graben** (a German term meaning "trough"), or a **rift valley.** There are a number of excellent examples of this landform, such as California's Death Valley, the Rhine Graben, which extends from Basel, Switzerland, to Mainz, Germany, and the rift zone that stretches northward from Lake Nyasa in East Africa through Lake Tanganyika, the trough of the Red Sea, and the valley of the River Jordan in Israel.

When you think of the great age of this planet of ours and observe such evidence of past diastrophic forces as elevated coastlines, great mountain ranges, as well as

FIG. 10–5 Faults. (on facing page) (a) illustrates the terminology associated with faults. (b) shows a normal fault, (c) a reverse fault, (d) a strike-slip fault, (e) a fault-block uplift, also called a horst, and (f) a fault-block valley, called a graben.

(a)

(b)

(c)

(d)

(e)

(f)

such common sights as faults and folds exposed in a highway cut, you may think that all of these things are from the past. When you consider how many millions of years these things have been happening, it is difficult to relate such processes to our own time. However, diastrophism is a continuing process, as the prevalence of earthquakes worldwide conclusively proves (Fig. 10–6).

Most earthquakes are noticed only by an instrument called a **seismograph,** which indicates the intensity of the earth's vibrations.* Some earthquakes, however, are quite violent, resulting in considerable damage to property and even to human life. Severe earthquakes have occurred all over the world, as in Tashkent, Soviet Central Asia, in 1966 when 75,000 people were left homeless; Managua, Nicaragua, December 1972, 5000 killed, 20,000 injured, 250,000 left homeless; Fars, Iran, April 1972, 5000 killed; Lisbon, Portugal, in the year 1755, 50,000 killed; Yokohama and Tokyo, 1922, over 150,000 killed; and San Francisco, California, 1906, 100 killed and much of the city destroyed.†

The question always arises after every serious earthquake: Is there any way to predict or forecast an earthquake? Surely, there must be some way to warn the citizens of a village, town, or city in an active earthquake region that a seismic wave (from the Greek *seio,* "to shake") is on the way, as they do during the tornado season throughout the American Midwest. **Seismologists** have been working on this problem for a long time. They have produced instruments (seismographs) to locate the epicenters (central points) of earthquakes and scales (sets of conversion tables) to determine the magnitude of the quake (such as the **Richter scale,** which indicates the actual amount of energy involved, and the **Mercalli scale,** which describes the local effect or damage caused by an earthquake). They have not as yet, however, solved the problem of earthquake prediction.

The nature of the problem is complex. Not all earthquakes have the same origin. Some are primarily horizontal, sending great waves traveling through the earth's crust in a straight line, causing tall buildings to sway dangerously. Others have a rotary motion, and still others are short lived, with a vertical thrust that makes the ground seem to jump up and down. Some occur within a landmass; others occur on the ocean floor, usually near the edges of the earth's plates.

Some seismologists believe that the **elastic-rebound theory** best explains how earthquakes occur. Rock layers beneath the earth's surface are subjected to intense pressures from many angles. At first they bend, then they snap, as the relieved sections try to return to their original position. Seismic waves are sent rolling through the earth's crust, and an earthquake results.

---

*The vibrations are of two basic types, a compression wave and a transverse or shear wave. Inasmuch as the compression waves travel faster through the earth, they arrive at a distant point ahead of the transverse waves, and thus are known as primary or P waves. The transverse waves arriving later are referred to as secondary or S waves. If one witnesses an earthquake, he will possibly note first a sharp thud, or blastlike shock, which marks the arrival of the P wave, then a few seconds later a swaying or rolling motion may be felt, which marks the arrival of the S waves.[5]

†To learn more about the earth's devastating earthquakes, see the United States Department of the Interior, Geological Survey *Earthquake Information Bulletin,* published by the Government Printing Office, bimonthly; and William C. Putnam (1964), "Earthquakes and the Earth's Interior," in *Geology* (New York: Oxford University Press), chap. 9.

Earthquake zones—regions of highest seismic activity

FIG. 10-6 Earthquake Zones of the World. (Adapted from U.S. Department of Commerce, Earthquake Information Bulletin, 5, no. 2, 1971.)

When an earthquake occurs in the floor of an ocean, the shaking of the sea bottom sets waves in motion, which then travel great distances. These seismic sea waves have inflicted great damage to coastal cities, from Hawaii to Japan. They are popularly known as **tidal waves,** although the pull of the moon has nothing to do with their origin. The more proper term is **tsunami** (which comes to us from the Japanese, who have endured more than their share of seismic sea waves). Hilo, Hawaii, has suffered several tsunamis, the most notable one occurring in 1960. The city received word that a tsunami was going to strike the waterfront, but despite radio warnings to move to high land, tourists ignored the advice and rushed down to the beaches to watch the "tidal waves" come in. Many were drowned as the 20- to 30-foot (6- to 9-meter) waves rolled into the low-lying city and poured into second-story windows of the buildings.

The search for answers is continuing. A theory now increasingly accepted suggests that before earthquakes strike they always send advance notice in the form of gradually increasing shock waves. Proponents of this theory believe that a warning system of computer-connected instruments could be set up along active fault lines in earthquake regions. If, after a period of quiet, a tiny shock agitates the instrument, then is followed by a growing stream of more intense shocks, the seismologist would know that an earthquake was in the offing, and he might be able to predict the time, place, and intensity of the quake.* Soviet and Japanese seismologists are also working on this problem and have indicated a willingness to share their discoveries with us.

Prediction is one thing; prevention is another. Some seismologists are trying to determine ways to eliminate earthquakes before they cause any damage. The idea is to lubricate the fault line by injecting oil or water into the rift or fracture zone. Presumably, the fault blocks will then slide smoothly rather than jerking violently. People who reside in earthquake territory hope that these theories prove to be practical.

VULCANISM   Volcanic activity can be quite picturesque—if you are a tourist visiting the Hawaii Volcanoes National Park, where you can drive to within 100 yards (90 meters) of a steaming lava flow. You stand impressed and safe under the watchful eyes of a park ranger, knowing that the entire island was created out of a series of such flows over millions of years of time. But volcanic activity can be terrifying if you are forced from your home by a flaming wall of lava, as happened to the citizens of Heimaey, a volcanic island that is part of the Mid-Atlantic Ridge, just south of Iceland.

Earth's history records a number of extremely violent volcanic eruptions. In A.D. 79 Mt. Vesuvius, Italy, destroyed the city of Pompeii; in 1902, Mt. Pelée, on the island of Martinique, the West Indies, leveled the town of St. Pierre; in 1883, a volcanic island called Krakatoa, which is west of Java, erupted, sending a violent tsunami against nearby islands, causing the deaths of almost 40,000 people; and Mt. Etna on the island of Sicily has erupted repeatedly (about 400 times since 475 B.C.),

*To learn more about the possibility of earthquake prediction, see Christopher Scholz et al. (31 August 1973), "Earthquake Prediction: A Physical Basis," *Science,* 181, no. 4102, 803–809.

killing almost 15,000 people in 1169 and 20,000 in 1669.[6] In February 1943, volcanic activity erupted on a farm near the Mexican village of Parícutin. Not only was the farm lost, but the village as well, as the earth belched forth fire and steam, sending great clouds of volcanic ash into the sky. Where the farm once was, a 1500-foot (450-meter) volcano now stands.

The causes of **vulcanism** are not known for certain. In our earlier discussion of the different theories regarding tectonic forces, the most likely to account for worldwide volcanic activity was the theory of plate tectonics, which included continental drift and sea-floor spreading. Figure 10–7 shows the distribution of the world's volcanoes, and it is obvious that most of them occur along the edges of the earth's plates. Because the Pacific Ocean is completely ringed with volcanoes, this collection is often termed the Pacific "ring of fire."

Figure 10–8 presents several cross sections of the sea bottom. Figure 10–8(a) focuses on the North Atlantic, showing how the ocean floor is separating along the fracture line called the Mid-Atlantic Ridge. As the American and Eurasian plates move apart, magma flows upward, creating volcanic mountains (islands), such as the Azores and

Eruption!   The village of Vestmannaeyjar on the volcanic island of Heimaey is overwhelmed by the volcano Helgafel. Heimaey, part of the Mid-Atlantic Ridge, is south of Iceland.

FIG. 10-7 Distribution of World's Volcanoes.

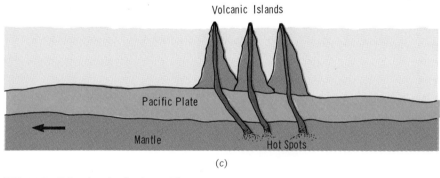

FIG. 10-8 Vulcanism in the Ocean Floor.

the Canary Islands. Figure 10–8(*b*) is of the western edge of the North Pacific Ocean and illustrates how the Pacific plate plunges down into the mantle near the coastline of Asia, forming deep trenches (deeps) and allowing magma to rise and form volcanic islands such as the Marianas. Figure 10–8(*c*) shows how the movement of the Pacific plate over the mantle creates "hot spots" because of the friction, with magma rising through points of weakness to build volcanic islands such as the Hawaiian Islands and the Tuamotu Archipelago.

There are two basic types of volcanic activity: intrusive and extrusive. **Intrusive vulcanism** (Fig. 10–9) results when magma cannot find an outlet from the asthenosphere onto the earth's surface. If its way is blocked, the magma may ooze through

(a)

(b)

(c)

Magma ▢ Strata, or layers, of sedimentary
or metamorphic rock

FIG. 10-9 Intrusive Vulcanism. In (a), a drawing of exposed batholith, overlying layers, or strata, have been eroded away, exposing the granitic mass of the batholith. (b) shows smaller intrusions, including a laccolith, dike, and sill. The neck or vent of a volcano is exposed after the overlying basaltic lava and cinders have been eroded away (c).

The Rocky Mountains: Exposed Batholith.

cracks and fissures and send tiny fingers or broad sheets between the layers (strata) of sedimentary rocks. These slablike intrusions are called **sills**. If the magma rises through the beds of rocks, cutting across the layers, the result is a **dike**. The large chamber of magma deep beneath the surface is called a **batholith** (from the Greek *bathys,* "deep," and *lithos,* "rock"). If the chamber of magma lies close to the surface and forces the rock layers above to arch upward, creating a structural dome, the result is a **laccolith** (from the Greek *lakkos,* "pit"). The main difference between a laccolith and a batholith is size (one might say that a laccolith is a small batholith). Some batholiths extend for great distances, such as the Rocky Mountains. In time, the overlying sedimentary rocks are worn away, thus exposing the vast expanse of basaltic rocks to view. Another exciting landform created by intrusive vulcanism is New Mexico's Ship Rock, an exposed network of dikes surrounding the stock (or neck) of an ancient volcano.

Extrusive vulcanism occurs when magma issues forth from the neck of a volcano, either in liquid form as **lava**, flowing like a river, or as **ejecta**, with volcanic ashes or pyroclastic bombs being violently ejected. (**Pyroclastic** means "fragments of rocks born of fire.") The landforms produced (Fig. 10–10) range from volcanic cones (Fig. 10–10[a]) to broad lava plateaus (Fig. 10–10[c]). The Columbian Plateau in Idaho covers a great area of the American Northwest, and one of the most interesting sections is Craters of the Moon National Monument. Astronauts who have returned from a landing on the moon say that this region is strikingly similar to the surface of the moon, complete with the glassy, black sand and numerous craters.

Volcanic eruptions can take many forms. If a developing hot spot happens to be situated near a large body of water, the magma may take in considerable moisture and thus become more fluidlike in consistency. When the magma is extruded from its chamber, it will flow smoothly out onto the surface. Gases in the magma tend to dissipate quickly, along with the water, in great clouds of steam. If the lava is fluid enough, the outpourings will be smooth surfaced, much like the flow of chocolate syrup down the slope of a mound of ice cream. If this type of flow occurs

Craters of the Moon National Monument, Idaho.

Crater

Crater

Magma

(a)

Crater

Crater

(b)

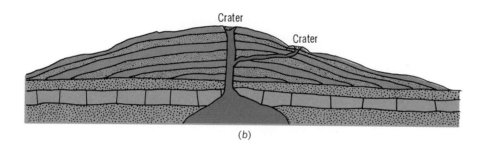

Strata of Sedimentary or Metamorphic Rock

Sill

Dike

(c)

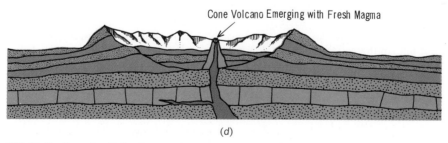

Cone Volcano Emerging with Fresh Magma

(d)

FIG. 10–10 Extrusive Vulcanism. (a) cinder-cone volcano; (b) dome, or shield, volcano; (c) lava plateau; (d) caldera, an enlarged crater that results when the top of a volcano either blows its top off or collapses downward.

repeatedly, the resulting volcanic mountain will be shaped like a shield or dome. A good example of a **shield volcano** is Mauna Loa of Hawaii.

If the eruption comes in the form of a blast because its pipe (neck or vent) is blocked, peices of rock and ashes will be flung high into the air. Subsequent similar eruptions will build a steep-sided **cinder-cone volcano,** such as Vesuvius of Italy. If there are alternating eruptions, some fluidlike, some pyroclastic, the volcano will have layers of ashes and ejecta, layers of lava, and the result is a **composite volcano.** Good examples of this type of volcano are Stromboli of the Mediterranean, Fuji of Japan, Mount Lassen in the United States.

When a volcano ceases to be active—that is, when there are no signs of activity over a long period of time—it is said to be dormant. We do not like to use the term *dead,* for who knows when it may suddenly return to life and start erupting again?

Other activities associated with vulcanism include geysers, hot springs, mud pots, and fumaroles. These result when water seeps down through cracks or fissures of the earth's crust and approaches close to the mantle or a chamber of magma. When the water is heated to extreme temperatures, it rises. Some of it is forced back out of the fissure as bubbling water, the remainder explodes upward as steam, spouting in great clouds above the earth. Geysers are the most popular attractions at Yellowstone National Park. Other countries noted for their geysers are Iceland, Japan, and New Zealand, all located on the edge of crustal plates and surrounded by oceans.

One final note about vulcanism. A United Press International news report from Hilo, Hawaii, dated March 18, 1973, tells of a circuit court judge's ruling that "any new land formed by a lava flow becomes the property of the owner of adjacent land." Thus ended a legal battle that resulted from a 1955 lava flow from Kilauea Iki.

Landform Classification

In order to assess the importance of the earth's many landforms, we should consider their individual characteristics. Figure 10–11 is one method of classifying the surface features of land and sea. We have divided the features into mountains, hills, plateaus, plains, and trenches (see again Fig. 10–3). Factors to be considered are: local relief, elevation above or below sea level, and the physical appearance.

MOUNTAINS AND HILLS It is sometimes difficult to distinguish between mountain and hill landforms. Mountains are supposed to be rough and craggy, with a **local relief** (the difference between the highest elevations and the lowest in a given area) of at least 2000 feet (600 meters) and with an average elevation of over 3000 to 4000 feet (900 to 1200 meters). Hills, on the other hand, are supposed to have a local relief of from 500 to 1000 feet (150 to 300 meters), with an average elevation of less than 2000 feet (600 meters). In some regions, however, local inhabitants refer to hilly country in their locality as "mountains," even though the hills may barely top 2000 feet (600 meters). Some of the so-called mountains in the Los Angeles area, such as the Santa Monica Mountains, do not meet the specifications for height

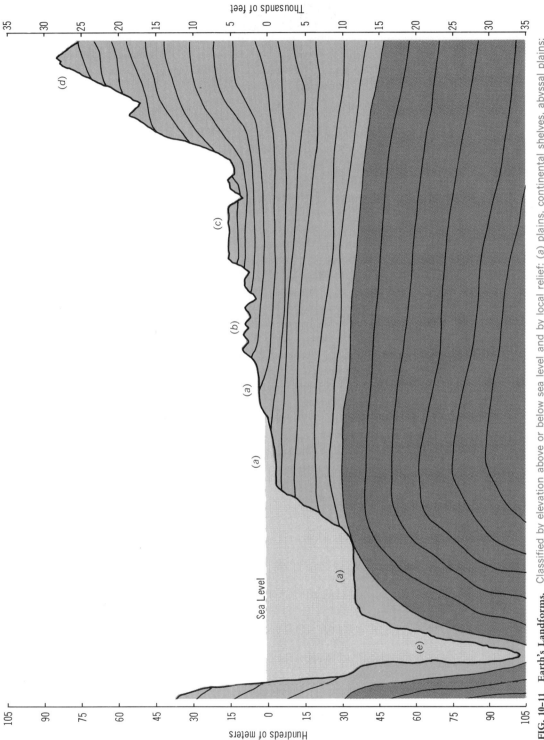

FIG. 10-11   Earth's Landforms.   Classified by elevation above or below sea level and by local relief: (a) plains, continental shelves, abyssal plains; (b) hills; (c) plateaus; (d) mountains; (e) trenches.

or ruggedness, and the elevations do not exceed 1800 feet (540 meters). Hills are usually smoothly rounded and can be formed by any of the tectonic forces (diastrophism or vulcanism) we have discussed. These are termed *structural hills.* In addition, they can be carved out of flat land by running water or glaciers. In chapter 15, we shall discuss how glaciers can also deposit small mounds of sediments that create a fertile, though hilly, landscape.

PLAINS AND PLATEAUS   Both plains and plateaus are considered to be relatively flat lands, with a local relief of less than 500 feet (150 meters). The prime difference between the two is height above sea level. Plains are usually found at elevations of less than 500 feet (150 meters), while plateaus may reach higher altitudes, up to 10,000 feet (3000 meters). Some earth scientists refer to plateaus as "plains in the air," or "tablelands." There normally are two classifications of plains. The first is **depositional,** which is made up of stream- or wind-deposited soils. Examples are delta (mouth of a river), flood (valley of a river), and **piedmont** (along the base of a mountain range). The second classification is **erosional,** where hilly or mountainous land has been worn away by stream, wind, or glacial actions. Plateaus are formed in the same ways as plains; thus their classifications focus on the activities that formed them. These classifications are: **intermontane** (located between two mountain ranges), piedmont, **continental** (covering vast expanses, such as North Africa), and **dissected** (deeply cut by a powerful river, such as the Colorado Plateau).

Most of the great plateau regions of the world are to be found on the **continental shields** (Fig. 10–12). These are exposed bedrock areas, lands composed of granite and some metamorphic rocks that have been subjected to extreme glacial action. They were initially formed some 3 billion years ago and experienced tropical ages, with great forests and very warm climates. With the passage of millions of years, the climate of the world changed many times. Great ice ages occurred, sending vast sheets of ice (continental glaciers) moving equatorward from the poles. The ice scraped away any hilly country, leaving the granitic masses exposed. Every continent contains at least one of these shields as vast plains or plateaus, but time and the elements have treated them differently, and they are not all alike today. As climate varies with latitude and position relative to the oceans and wind systems, the rock shields have not evolved at the same rate. The climate, soils, and vegetation of the Canadian and Angara shields cannot be compared to those of the Brazilian or African shields.

Another great plains region can be found on the floor of the ocean. The deep-lying plains are called **abyssal plains.** The outer borders of the continents, the edge of land that extends into the oceans, contains the continental shelves. We shall consider these in greater detail in chapter 15, but for now let us state that these shelves are important to mankind because of the life forms that thrive in the shallow waters and the great petroleum deposits beneath the surface of the shelves.

LANDFORMS OF THE UNITED STATES AND CANADA   One of the most common ways devised by geomorphologists to explain and portray the landforms of the United States and Canada is the one shown in Fig. 10–13. This physiographic map shows the major landform features we have been discussing. You will note that some of

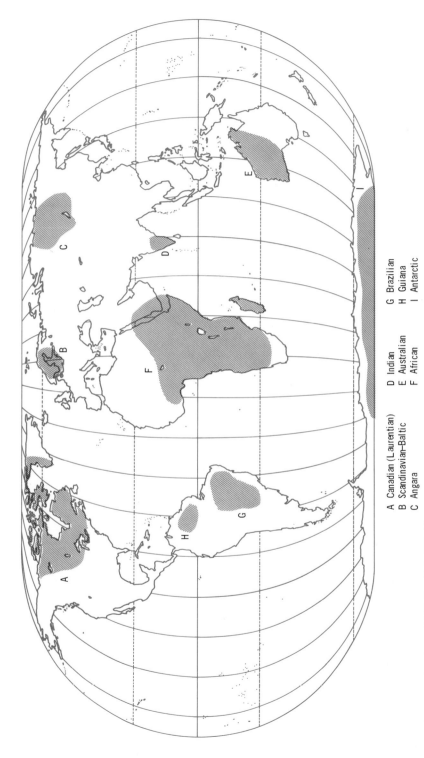

FIG. 10-12 Continental Shields of the World.

A Canadian (Laurentian)
B Scandinavian–Baltic
C Angara
D Indian
E Australian
F African
G Brazilian
H Guiana
I Antarctic

1 Hudson Bay Lowland
2 Rocky Mountains
3 Canadian Shield (Laurentian Upland)
4 Great Plains
5 Appalachian Uplands and Mountains
6 Columbia Plateau
7 Pacific (coastal) ranges

8 Central Plains (lowlands)
9 Great Basin (and range)
  Intermontane Plateau
10 Colorado Plateau
11 Interior uplands
12 Gulf coastal plain
13 Atlantic coastal plain

Fig. 10-13 Landforms of the United States and Canada.

the terms are not the same as those we have used. These different terms are employed to further clarify the differences between landforms that have similar characteristics. **Uplands,** for example, is a term referring to hilly country or high plains areas that do not qualify as plateaus. **Lowlands** are low-lying lands that are fairly close to sea level in elevation and normally swampy or dotted with streams, rivers, and lakes. The term **basin and range** indicates a relatively flat plain or high plain region that is dotted with isolated mountains or ranges.

Mountain areas in Fig. 10–13 are indicated by the numbers 2, 5, and 7; hill country, 11; plateaus, 6, 9, and 10; plains, 1, 3, 4, 8, 12, and 13.

Review and Discussion

1. Explain the term *tectonic*. What are tectonic forces?

2. Explain the difference between these terms: *contraction, convection.*

3. Explain the theory of continental drift. What was Pangaea?

4. How did a discovery made by scientists in the field of paleomagnetism help to make the theory of continental drift more acceptable?

5. What is the meaning of isostasy?

6. What does the Pacific plate have to do with California's San Andreas fault?

7. Explain the difference between epeirogeny and orogeny.

8. Explain the difference between diastrophism and vulcanism.

9. Explain how folding occurs. What is the difference between folding and warping? What is the difference between folding and faulting?

10. What is an anticline? What is a syncline?

11. Explain the differences between these fault types: normal, reverse, strike-slip.

12. What is the difference between a horst and a graben?

13. Where would you find the greatest concentration of earthquake activity?

14. What is the difference between the Richter scale and the Mercalli scale?

15. What is the difference between a tidal wave and a tsunami?

16. What is meant by the term *ring of fire?*

17. Explain the difference between intrusive and extrusive vulcanism.

18. How does a sea bottom trench or deep come into being?

19. What is the difference between a laccolith and a batholith?

20. What is the difference between a sill and a dike?

21. Define these terms: *ejecta, pyroclastic, lava, magma.*

22. What is the difference between cinder-cone, shield, and composite volcanoes?

23. What is meant by the term *local relief*, with reference to landform classification?

24. What are the major differences between hills and mountains, plains and plateaus?

25. Explain these terms that refer to plateaus: *intermontane, piedmont, dissected, lava.*

26. What is a continental shield?

27. What is the difference between abyssal plains and continental shelves?

Notes

[1] *Larousse Encyclopedia of Mythology* (1959), trans. Richard Aldington and Delano Ames (London: Paul Hamlyn), p. 423.

[2] *Larousse Encyclopedia,* p. 447.

[3] *Larousse Encyclopedia,* pp. 139–142.

[4] *Larousse Encyclopedia,* p. 218.

[5] U.S. Geological Survey (1973), *San Andreas Fault,* publication no. 0-495-648 (Washington, D.C.: U.S. Government Printing Office), p. 6.

[6] Leon Bertin (1972), *The New Larousse Encyclopedia of the Earth* (New York: Crown Publishers), p. 163.

Additional Reading

Anderson, Don L. (November 1971), "The San Andreas Fault," *Scientific American* (offprint no. 896).

Beiser, Arthur (1962). *The Earth.* New York: Time-Life Books.

Bloom, A. L. (1969). *The Surface of the Earth.* Englewood Cliffs, N.J.: Prentice-Hall.

Dalrymple, E. S., and E. Jackson (May–June 1973). "Origin of the Hawaiian Islands," *American Scientist,* 61, no. 3, 294–308.

Dury, G. H. (1960). *The Face of the Earth.* Baltimore, Md.: Penguin Books.

Du Toit, A. L. (1937). *Our Wandering Continents.* Edinburgh, Scotland: Oliver and Boyd.

Grove, Noel (July 1973). "A Village Fights for Its Life," *National Geographic,* 144, no. 1, 40–67.

Heirtzler, J. R. (December 1968). "Seafloor Spreading," *Scientific American* (offprint no. 875).

Herbert, Don, and Fulvio Bardossi (1968). *Kilauea: Case History of a Volcano.* New York: Harper & Row.

Hodgson, J. H. (1964). *Earthquakes and Earth Structure.* Englewood Cliffs, N.J.: Prentice-Hall.

Hurley, P. M. (April 1968). "The Confirmation of Continental Drift," *Scientific American* (offprint no. 874).

Iacopi, Robert (1971). *Earthquake Country.* Menlo Park, Calif.: Lane Books.

Kay, Marshall (September 1955). "The Origin of the Continents," *Scientific American* (offprint no. 816).

Kennedy, George C. (December 1959). "The Origin of Continents, Mountain Ranges, and Ocean Basins," *American Scientist,* 47, no. 4, 491–504.

Macdonald, G. A., and A. T. Abbott (1970). *Volcanoes in the Sea: The Geology of Hawaii.* Honolulu: University of Hawaii Press.

Matumoto, T., and Gary Latham (10 August 1973). "Aftershocks and Intensity of the Managua Earthquake of 23 December 1972," *Science,* 181, no. 4099, 545–547.

Milne, L. J., and Margery Milne (1962). *The Mountains.* New York: Time-Life Books. See especially, "Volcanoes," chap. 3, and "Birth and Death of Mountains," chap. 2.

Ogburn, Charlton, Jr. (1968). *The Forging of Our Continent.* New York: American Heritage.

Scholz, C., L. Sykes, and Y. Aggarwal (31 August 1973). "Earthquake Prediction: A Physical Basis," *Science,* 181, no. 4102, 803–809.

Shelton, John S. (1966). *Geology Illustrated.* San Francisco: Freeman.

Snead, Rodman (1971). *Atlas of World Physical Features*. New York: Wiley.

Tarling, Don, and Maureen Tarling (1971). *Continental Drift: A Study of the Earth's Moving Surface*. Garden City, N.Y.: Doubleday.

Tazieff, Haroun (1961). *The Orion Book of Volcanoes*. New York: Orion Press.

Wegener, Alfred (1929, 1966). *The Origin of Continents and Oceans*. New York: Dover Publications.

Wilson, J. Tuzo (April 1963). "Continental Drift," *Scientific American* (offprint no. 868).

The Earth's Sculptured Face

PART 4

11

HOW THE SCULPTURING BEGINS

The earth must endure endless cycles:
seasonal,
hydrologic,
vegetative,
geographical.

Geographical cycle?

What goes up, must come down . . .
Up . . . down . . . up . . . down . . .

There is no end . . .

If the term cycle means a "round of events that recur again and again," how can there be a geographical cycle? Does this mean that the earth's surface is constantly being uplifted and then worn down? For an answer, let us consider the broad picture. The earth's tectonic forces operate to raise some portions of the crust and depress others; continents are set adrift; mountains are uplifted; plateau regions are elevated; and the ocean floors are spread apart. The story does not end there, because other external (outside) forces are working to level the land. These forces are called *gradational* because they operate to grade down or wear away elevated surfaces in gradual stages. As rapidly as an area is pushed up, the gradational forces work to bring it down. Earth materials are first broken up, then removed from the higher places, transported to lower elevations, and deposited. One might say that the higher elevations are denuded (stripped of their cover) and the lower places aggraded (built up). When the lower place is uplifted, it will become denuded. This is, apparently, a continuous process, a true cycle. It is called, variously, the **cycle of denudation,** the **geomorphic cycle,** or the **geographical cycle.***

Our next step is to consider the agents working within the geographical cycle to sculpture the face of the earth, to level the land, and to allow the development of soils needed by mankind. We shall examine the erosional (loosening, wearing down, removal, and transportation) and depositional activities of the forces of gravity (mass wasting), running (surface) water, groundwater, glaciers, ocean waves and currents, and wind. First, however, we must learn how the bedrock, the parent material, is broken up and loosened, because weathering is the first step is this process.

Weathering

Have you ever wondered why the paint on your house peels and flakes away? Or have you looked at a car you bought five years ago and wondered why the paint looked different than when it was new? Or maybe you have stood in front of a towering granite cliff and noticed that the face of the cliff was slowly breaking down and was not as slick and shiny as you remembered it from when you camped in the area ten years ago.

* To learn more about this and opposing views, see: William Warntz and Peter Wolff (1971), "Theoretical Models for the Study of Landforms: Davis and Horton," *Breakthroughs in Geography* (New York: New American Library), chap. 7; Kirk Bryan (December 1940), "The Retreat of Slopes," *Annals of the Association of American Geographers,* 30, no. 4, 254–268.

All of these things can be attributed to **weathering**. A number of factors are involved, such as the action of the weather—the amount of insolation received (based on the latitude, elevation, and aspect), the annual and daily ranges of temperature, and the amount and kind of precipitation. In addition, the activities of burrowing animals, bacteria, and the roots of plants are important weathering agents in breaking up rock materials. The process of weathering includes both chemical and mechanical (or physical) actions.

MECHANICAL WEATHERING   When rock masses are fractured and broken, we call the process **disintegration,** which means the rock falls and splits or crumbles into small pieces, grains, or particles. Disintegration is strictly a physical action, as there is no chemical change within the rock structure. If you were to examine a chunk of granite and compare its chemical makeup with a handful of California beach sand, you would find that they have exactly the same mineral combination: feldspar, mica, and quartz. The sand was granite before it was physically weathered. Rock layers within the earth's crust and upper mantle can be fractured (jointed) by tectonic forces, and if the overlying materials were removed by erosion, these once-buried masses would be subjected to further weathering.

Rocks can be fragmented by a number of processes. One such process is termed **frost action,** where water seeps into rock crevices and expands as it freezes, causing the rock to break apart along the fracture lines. Sometimes the moisture in the rock develops ice crystals as the temperature descends below freezing. The growth of these crystals exerts pressure on the rock mass from within, and the rock begins to break up. This process, termed **crystal growth,** also occurs in areas where the rate of evaporation is high and the heat draws moisture to the surface. This upward-moving moisture contains minerals it has dissolved out of the lower regions of the rock. On reaching the surface, the moisture rises through evaporation, leaving the chemicals (such as salt) behind, lodged in crevices of the rock. If the salt crystals increase in number and in size, the pressures may force the rock to split apart. Frost action and the formation of ice crystals are common occurrences in the high and middle latitude regions, while the action of salt crystal growth is more prevalent in the lower latitudes and in dry regions.

Another physical weathering process involves the alternating of high and low temperatures, which causes rocks on or near the surface to expand and contract repeatedly, thus weakening the rock structure. This is called **thermal expansion,** or **thermal weathering.** For example, consider the effect of extremes of temperature on a large boulder. The outer portion being heated during the day expands, while the interior, receiving little or no heat, does not expand. When night falls, the cooler air temperature chills the outer rock structure, causing contraction. The interior is not affected. The repeated expansion and contraction causes the rock to break up from the outside.

Another type of physical breakup is caused by plant roots working their way down into the joints and crevices of a rock layer. The roots expand as they grow, pushing against the sides of the crevice and forcing the rock to break apart. Tiny burrowing animals contribute to rock breakup by working their way to and fro through cracks in the rock, and their constant passage wears away the rock surfaces. This type of weathering is termed **biotic**, or **organic**, **activity**, or the work of living organisms.

Rocks can also be worn down by the abrasive (sandpapering) action of running water and winds, the plucking (removal) of rock fragments by moving glaciers, and by a process called **unloading**. When a rock mass is buried under other layers, it is compressed by the resulting pressures. If the overlying rock layers are eroded away, the rock layer may expand, and as it does it will fracture and break apart.

What about the human element? Think of the effect of our technology on the rock structures: automobiles, motorcycles, trucks, bulldozers, and all kinds of earth-moving equipment that travel the roads and highways of the world. Do you think that the weight of these vehicles has any effect on rocks underneath? What of high-flying jet planes? If sonic booms can shatter windows, can they also fracture rocks? We might term this type of physical weathering cultural or mankind induced.

CHEMICAL WEATHERING When rock breakup is brought about because of alterations in the chemical composition of a rock, the result is termed **decomposition** (rotting). Sometimes chemicals are dissolved by gases or moisture and carried away, thus leaving greater spaces between rock particles or grains. The weakened rock is left open to further decomposition, as well as to disintegration.

If water enters a rock layer containing carbon dioxide, a mild form of carbonic acid is formed that dissolves some existing chemicals within the rock, such as limestone (calcium carbonate). This process, **carbonation**, has created numerous underground caverns in limestone regions, such as Mammoth Caves, Kentucky. If oxygen is present, rusting (**oxidation**) will occur.

When moisture creeps into rock cavities, it may unite with any mineral present, thus causing chemical changes. If this happens, the rock may break up, an action called **hydrolysis**. When the rock absorbs moisture but there is no resulting chemical alteration, the rock may expand in size, and if so, it will rupture. This is called **hydration**.

If a rock is weakened by the addition of moisture, which dissolves the chemicals that cement the rock grains or particles together, the rock will decompose as the binding cement is carried away. Some rocks are composed entirely of easily dissolved minerals, and water will bring rapid decomposition. This is similar to what would happen if you were to patch a broken ceramic vase with a water-soluble glue and then put the vase in a tub of water. The glue would quickly dissolve, and the vase would fall apart. This process is called **solution**.

THE RESULTS OF WEATHERING The basic rock structure of the earth is called *bedrock*. When weathering alters the bedrock by breaking it into smaller and smaller pieces, it eventually becomes **regolith** (from Greek *rhegos,* "blanket," and *lithos,* "rock"). This "blanket of rock" is the unconsolidated (loose) rock material that lies above the bedrock. Regolith in one place may not match that of another, as there are many factors contributing to the eventual shapes of the rock fragments. Depending on the chemical composition, granular structure, or the climate of an

(a)

(b)

(c)

(d)

(e)

Weathering, Erosion, and Deposition: Sculpturing Agents. (a) The combined efforts of sun, wind, and rain have carved the fantastic columns of Bryce Canyon National Park, Utah. (b) Wind-driven waves pound the rocky coastline of St. Andrews State Park, Florida. (c) Thundering waterfall along the Pigeon River (on the Minnesota-Ontario border) easily shatters and fractures highly resistant beds of rock. (d) Streams flowing over gently sloping ground in Wisconsin's cattle country carve deep gullies in unprotected land. (e) Silt transported by a hurricane overloaded this normally clear running stream, causing heavy deposition of the silt on both sides of the stream's valley.

314

The Results of Weathering. (a) granular disintegration of boulders, Joshua Tree National Monument, California; (b) joint-block separation of a quartzite cliff in Bennington, Vermont; (c) spalling (exfoliation) of granite domes in Yosemite National Park, California; (d) shattering of an ancient basaltic lava flow in Grand Coulee, Oregon.

(a)

(b)

(c)

(d)

(a)

Mass Wasting: Transport of Earth Materials by the Pull of Gravity. (a) massive earthflow (caused by water seepage high on the hillside) that covered U.S. 10 and the Cascade Irrigation Canal in Ellensburg, Washington; (b) mudflow in Colorado that has been in existence for more than 100 years, and flows 20 feet (6 meters) per year; (c) landslide in oil-shale cliffs, northeastern Colorado; (d) solifluction in the Seward Peninsula region of Alaska; when spring arrives, the top few inches of the permafrost thaw, sending lobes of soil and rock sliding down the slopes; (e) slump along Otter Creek, Vermont, with a vertical drop of about 30 feet (9 meters).

(b)

(c)

(d)

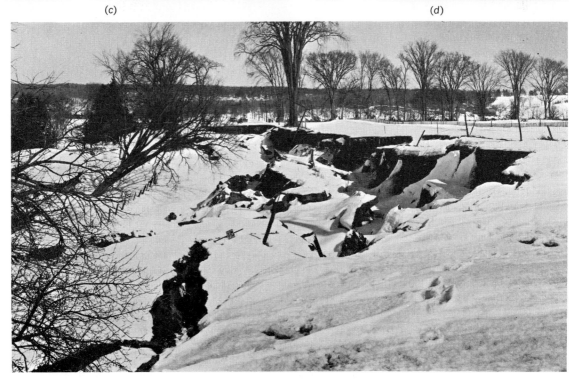

(e)

area, a number of different, but easily recognizable, shapes will result: small grains, curved shells, block- or bricklike pieces, or irregular chunks.

When rocks break up into irregular, sharp-edged, random-sized pieces, because they do not have definite lines of weakness, cleavage, or fracture, we say the rocks have **shattered.** If the rock separates along previously induced joints, the result looks like huge steps leading upward, and this is called **joint-block separation. Exfoliation** (from the Latin *exfoliatus,* "to strip of leaves") is the separation, through expansion, of curved shells or slabs from a solid rock mass that is uniform in structure. Excellent examples may be seen in the rounded domes of Yosemite National Park. Another term for this process is **spalling** (from the Old German *spellen,* "to split"). When a rock mass is composed of rocks with large crystalline structure, such as granite or sandstone, the rock may break up one grain after another, a process called **granular disintegration.**

The processes of physical and chemical weathering are important, but the breakup of rock masses does not necessarily produce landforms. The actual sculpturing of the earth's surface is carried on by the erosional activities of running (surface) water, groundwater, glaciers, ocean waves and currents, winds, and the force of gravity. This latter erosional activity is often ignored, but anyone who has lived through a landslide or whose car has been struck by a rock falling from a cliff knows that the erosional agent termed **mass wasting** can be an effective sculpturing agent.

Mass Wasting

The attraction of the earth upon any mass is called gravity, and there are a number of ways that this important erosional activity induces the removal of weathered rock materials: soil creep, earthflows, mudflows, rockfalls, landslides, and slumps. The degree of the wasting away of a mass depends on the volume of the mass and its weight, the degree of slope, and the amount of moisture it contains.

The gravity movement called **soil creep** occurs in areas of moderate-to-gentle slopes and can be caused by a deer moving down a trail denuded of grass cover, or by a beetle attempting to climb a sandy incline. **Earthflows** are common in areas of heavy precipitation where water-logged soils are commonplace. If the slopes are relatively steep, the heavy soils may slide like a great sheet of mud over wide areas. In a sense, the face of the bedrock under the surface is like a metal slide in a children's playground.

An interesting form of earthflow occurs in the tundra regions of Canada and Siberia. In this vast region of permanently frozen ground (permafrost), the short summers allow only the top few inches of the frozen soil, or regolith, to thaw out. The frozen ground underneath acts as a slide, allowing the thick, souplike top layer to flow slowly down the slopes. This process is termed **solifluction** (from the Latin *sol,* "sun," and *fluctus,* "wave"). This thawing of the permafrost causes serious problems for anyone constructing a building. Without a layer of insulation, the heat radiating down from the floor of the house will thaw the uppermost layer of the permafrost, and if there is a slope, the house may slide. If the land is level, a portion of the house may sink down into the ooze. It is also difficult to maintain highways, even if paved, in permafrost areas, because when the summer thaws begin, the roads buckle.

Mudflows, similar to earthflows, are common in semiarid regions. They occur after violent thunderstorms or heavy rain showers have saturated the soil in the topmost layers. Pulled by gravity, the mud begins to flow down the slopes. If the rains continue, the flow of mud will become faster. It will sweep down slopes, pour over or knock down fences and retaining walls, enter through doors and windows of homes, and sweep out onto a valley or plain beyond. Mudflows can also result from excessive water seepage from glaciers high in the mountains.

Landslides are probably the most dramatic of all the various forms of mass wasting. In many cases there is no advance warning as a large rock mass composed of regolith and soils suddenly breaks loose and thunders down a steep slope, perhaps to inundate a highway, mountain cabin, or an entire village. This occurs most often in high mountain areas such as the Alps, Rockies, and Andes. When a steep-faced

Landslide. This photo gives an idea of the devastating effects of landslides on man-made features such as this railroad in Grand Valley, Colorado.

cliff is undermined by erosional forces, the base may no longer be able to support the mass of rocks behind. Fractures result as the base begins to move; then the entire mass slumps down, giving us the term **slumping.**

Another action, called **rockfall,** results when the earth movement is localized and the mass being wasted is composed of individual pieces or fragments of regolith that break away from the edge or face of a cliff and fall tumbling to the valley floor below. When the pieces of broken rock build up sufficiently at the base, a **talus slope** comes into existence. Sometimes the fragments construct a cone-shaped structure called a **talus cone,** a landform that can be very difficult to climb.

Mass wasting is only one of the many forms of erosion, the forces that pick up weathered material and transport it to lower elevations, but all are important because they exert significant influences on the development of soils. And what would we do if we had no soils?

Review and Discussion

1. Explain the term *weathering.*

2. What are the differences between rock disintegration and decomposition?

3. Explain mechanical weathering.

4. What is meant by *frost action*?

5. What have temperature extremes to do with the breakup of rock?

6. Explain the term *crystal growth.*

7. Explain how small burrowing animals and microorganisms can contribute to weathering.

8. Can people be included in the weathering processes?

9. Explain the term *unloading.*

10. One of the processes of chemical weathering is *carbonation.* What does that term mean?

11. How does oxidation work to cause rock breakup?

12. Explain the difference between the terms *hydration* and *hydrolysis.*

13. Explain the term *solution* with regard to rock breakup.

14. What is regolith?

15. Explain the differences between: shattering, joint-block separation, spalling, and granular disintegration.

16. What does *mass wasting* mean?

17. Explain soil creep.

18. What are the differences between earthflows and mudflows?

19. What is permafrost?

20. What is solifluction? Where is it most likely to occur?

21. Explain the difference between landslides and slumps.

22. What is a talus cone?

Additional Reading

Putnam, William C. (1964). *Geology.* New York: Oxford University Press.

Reich, Parry (1950). *A Survey of Weathering Processes and Products.* Albuquerque: University of New Mexico Press.

Sharpe, C. F. (1938). *Landslides and Related Phenomena.* New York: Columbia University Press.

Shelton, John S. (1966). *Geology Illustrated.* San Francisco: W. H. Freeman.

Thornbury, William D. (1969). *Principles of Geomorphology.* 2d ed. New York: Wiley.

——— (1965). *Regional Geomorphology of the United States.* New York: Wiley.

12

The Evolution of Soils

A dictionary might define *soil* as something that is unclean; a filthy, soiling, substance; something dirty. Or, soil might be defined as the upper layer of the earth composed of loose surface material. Soil scientists, however, are more apt to state that soil is not merely a part of the lithosphere, but a very important and complex segment of the ecosystem, the biosphere: Soil is the layer of the earth that supports mankind.

The powerful forces of weathering and erosion operate to provide us with a physical environment that is both awe inspiring and beautiful, and they also produce the soils required for earth to maintain its vegetative cover. What is soil? It is a mixture of many substances: an almost endless variety of minerals, decomposing bodies of plants and animals, water, and air. When any of these predominates, a particular type of soil results. Because conditions differ from place to place, soils differ. Even the soils in your neighborhood differ from one area to the next. To be sure, soils in the Amazon rainforest region differ from those of Australia's Great Sandy Desert, but as well, soils from one section of Texas are unlike those from other sections.

Some soils extend downward to a depth of 2 or 3 feet (60 to 90 centimeters), while others are limited to a depth of only 2 or 3 inches (5 to 8 centimeters). Although some soils come into being almost instantly (as you will soon learn), most have evolved through several stages of development over hundreds or thousands or years.

Formation of Soils

Why do soils differ? **Pedologists** (soil scientists) have shown that soil formation is extremely complicated, so generalizations or narrow considerations, such as descriptions of the amount of precipitation and sunlight or the type of vegetative cover, are inadequate in discussions of soil variation. To understand the earth's soils, we must consider, first, the important controls that create different structures, textures, and fertility; second, the main processes involved in the gradual development of soils; and finally, the different methods of soil classification.

SOIL-FORMING CONTROLS   Pedologists say there are five major factors that control the formation of soils: climate, vegetation, time, parent material, and terrain.

Climate   The dominant soil-forming control is climate, because it is the source of sunlight and water. Precipitation in any of its forms can, like a double-edged sword, be beneficial or damaging. It can be beneficial when water from melting snow or falling rain penetrates the ground surface, and gravity causes this water to move to lower levels. This **gravitational water** takes with it chemicals, organic nutrients (humus), and fine soil particles, which become lodged in crevices of the regolith. It can be detrimental if one of the layers becomes heavily compacted with fine particles and, therefore, impervious to further downward movement of water, resulting in the formation of a *hardpan*. If the area's precipitation is light and the rate of evaporation high, moisture may be drawn upward toward the surface by a movement called **capillary action**. Chemicals left behind may create a thick layer of deposits that is as impermeable as clay-filled hardpans. If calcium carbonate is the chemical compound, as is common in the American Southwest, the result is called *caliche*.

In areas of heavy precipitation, chemicals and soil nutrients may be taken away as the gravitational water moves down through the soil and unites with groundwater flow. This "robbery" of needed nutrients is termed **leaching**, and areas where this occurs are notoriously infertile. Farmers in such areas must constantly move their field of operations from place to place, clearing new lands in their search for fertile soils. If a hardpan develops in an area of heavy precipitation, the layers of soil above the compacted layer may become waterlogged as the air spaces become filled with water. Except for hydrophytes (water-loving plants), plants may develop root diseases and die. As plants require oxygen to grow and perform their part in the formation of soils, the presence or absence of air is vital.

Corkscrew Swamp, Florida Everglades.   Waterlogged soils can support a thriving, luxuriant vegetation.

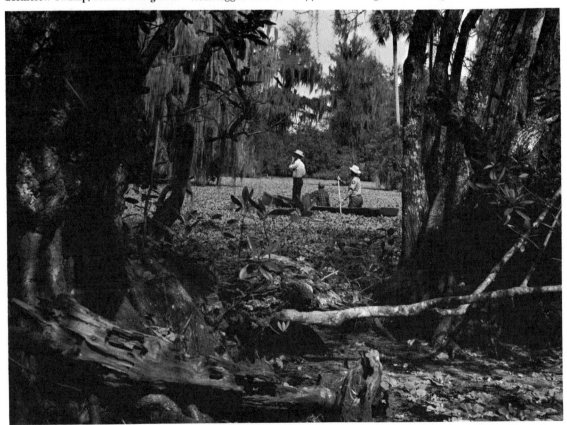

Temperature, whether consistently high or low, can encourage or discourage the proper evolution of a soil. High temperatures, if accompanied by sufficient precipitation, tend to increase the tempo of weathering activities. Bacterial activity speeds up, as does chemical decomposition, resulting in the accelerated breakup of the regolith. On the other hand, low temperatures tend to slow down the formation of soils, although physical weathering processes may be increased.

Vegetation Soil development is greatly influenced by vegetation. First, the roots of shrubs, grasses, and trees extend down into the soil structure as they reach for nutrients and moisture. They aerate the soil (supply oxygen and nitrogen), and they draw moisture from it. Second, the vegetation life forms act as a protective cover to hold the forming soil in place, and when they die, their bodies furnish humus to the soil. This is a continuing cycle: Vegetation requires humus so that it can grow and supply more humus.

Because conditions in grassland regions are favorable for humus formation and grasslands tend to retain humus near the surface, the grasslands of the world—for example, the steppes of Central Asia and the midwestern lowlands and the Great Plains of the United States—have the richest soils. What kind of soil would evolve in a desert region with low precipitation and high evaporation? What kind of vegetation would develop and what kind of humus could it contribute? Such soils might provide commercial minerals such as nitrates and gypsum for mining enterprises, but farmers would find them exceedingly unproductive.

Parent material The initial bedrock, the **parent material,** that is the source of a soil contains chemicals and mineral combinations, and if they are the right ones, fertile soil will evolve (assuming that other factors are favorable). Whether the soils have coarse or fine grains depends greatly on the original structure of the bedrock. Soil will evolve much faster from a sedimentary rock such as sandstone than it will from an igneous rock such as basaltic lava.

Time Another significant factor in the development of most soils is time. It may take thousands of years (under optimum conditions) for few inches of soil to develop from a block of solid granite or a formation of lava. What about soils found along rivers on in the deltas that form at their mouths? When deposits are left by a river that has overflowed its banks and spread its **alluvium** (stream-deposited material) on the land, will the resulting soil be fertile and ready for use? It can be if the original source was composed of fertile topsoils. Why do you suppose the ancient Egyptians referred to the River Nile as the "Holy Mother"? For thousands of years this river carried alluvium from the mountains of Ethiopia and the Ruwenzori of the eastern Congo through the winding channel every spring. When floodstage was reached, the river overflowed its banks and deposited the rich, fertile alluvium on the adjacent lowlands on either side.

Terrain The word **terrain** means the surface of the earth, whether it is smooth and low-lying, with gentle inclines, or rugged, hilly, or mountainous, with steep slopes. Flat or moderately sloping terrain is more likely to evolve good soil (if other factors are favorable as well) than steep slopes. Rugged, steeply sloping land allows

(a)

(b)

Influence of Time on the Development of Agriculturally Usable Soils. (a) relatively recent pahoehoe lava flow; (b) commercially profitable sugarcane plantation, in Hawaii, on soil that evolved from an ancient lava flow.

Fertile Soils along the River Nile, Egypt. (FAO Photo)

water to run off too extensively and too rapidly, so lowlands that are surrounded by higher land may tend to become saturated with too much moisture (waterlogged), thus forming swampy or marshy areas where soil may have a difficult time evolving. Here again, erosion is a factor. How can fertile soil evolve if it is always being depleted by violently running water or blowing winds? Often there is poor development of soil on mountain ridges, because the newly evolving soil is continually being carried away.

SOIL HORIZONS   So far in our discussion of soils, we have referred to soil layers and soil structure. Pedologists, however, are more specific; they use such terms as **soil profile** and **soil horizons** when focusing on specific soils and the differences between them. Each soil has its own structure or profile, and each profile is composed of layers or horizons. Figure 12–1 is a generalized view of a soil profile, showing the major horizons of its structure. The various soils of the earth have unique profiles, and each may not match any other. They also have horizons that vary considerably in thickness. In some, the topmost layer (the A horizon) may be only 3 inches (7.5 centimeters) thick, while not far away, out on a grassland, the horizon may well be 3 feet (0.9 meter) thick.

The **A horizon** is the top layer; its upper surface is the surface of the earth. In many places you cannot see any soil at all because of the litter (plants, leaves, branches) covering the surface. If bacteria are present, they will convert this material into

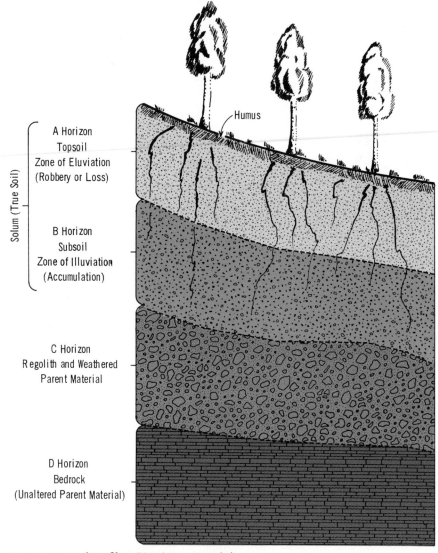

A Horizon
Topsoil
Zone of Eluviation
(Robbery or Loss)

B Horizon
Subsoil
Zone of Illuviation
(Accumulation)

Solum (True Soil)

Humus

C Horizon
Regolith and Weathered
Parent Material

D Horizon
Bedrock
(Unaltered Parent Material)

FIG. 12-1 A Soil Profile. (Not drawn to scale.)

decomposed matter, thus supplying humus to the soil. If moisture is present, this decaying material will be carried into the earth, and if there is a sufficient quantity, the A horizon will be quite fertile. This is not always true for A horizons, however, as may be seen in the soils of rainforest areas. Here, the horizon may be 20 feet (6 meters) thick but, nonetheless, infertile because of excessive leaching. The average thickness of the world's A horizons is about 7 inches (17.5 centimeters). Pedologists call this horizon the **zone of eluviation** (to wash out), but farmers call it topsoil.

The **B horizon** is normally a more compacted layer than the A horizon, because there is a heavier concentration of fine particles, such as silt or clay, brought down from above. This layer, therefore, is sometimes referred to as the **zone of accumulation.** Pedologists often use the term **zone of illuviation** (to move in). Because the air spaces have been filled with materials from the A horizon, water molecules have difficulty moving farther down through this horizon. In some regions, for one of two reasons this horizon may actually be richer than the upper layer: First, farmers may have used the land too heavily and not too wisely, thus robbing the A horizon of its fertility; and second, gravitational water may have caused excessive leaching

Well-Developed Soil Profile. Profiles with distinct horizons are characteristic of grasslands.

of the A horizon. In order to replenish the upper level, farmers sometimes must plow quite deeply to "turn up" this subsoil.

The **C horizon** is composed of weathered material (regolith) that is separating from the bedrock and has not evolved far enough to be considered a part of the true soil, or **solum** (a term often used by pedologists to refer to the top two horizons as a kind of separate body: the soil mantle). The C horizon may contain moisture as well as chemicals and nutrients, but it is not considered an agricultural soil.

The **D horizon** is the source of the soil evolving in the horizon above it (the C horizon). It may be solid, though fractured and jointed, or it may be fragmented in its topmost zone as it evolves and joins the regolith above it.

Horizon visibility can be quite clear, with the horizons having distinct boundaries, but sometimes it is extremely unclear, with blurred horizons, one zone seeming to

Blurred Soil Horizon.   Some soils have such indistinct horizons that it is difficult to construct a readable soil profile.

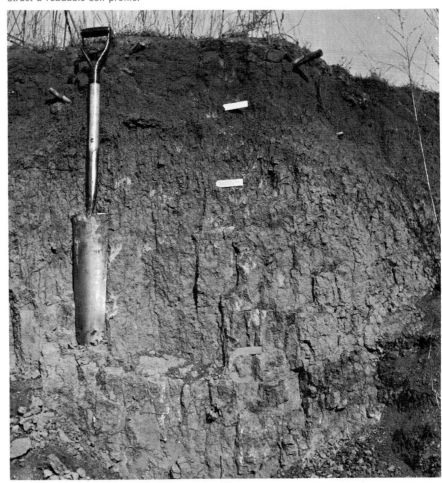

blend into the next. Alluvial soils seldom have definite horizons, because of the relatively short span of time it takes for them to evolve. Soils on mountain ridges or steep slopes may also have little or no A horizon, owing to heavy erosional action as the soil attempts to evolve.

DISTINGUISHING CHARACTERISTICS OF SOIL   A soil's color can aid in distinguishing its type. If there has been a heavy accumulation of humus on or near the surface, the A horizon will be dark, sometimes almost black. The famous steppes of the Ukraine are composed of dark soils, called *chernozem* (Russian for "black earth"). The dark color of these soils results from an abundance of humus in grassy areas. The color tends to lighten with increase in depth. The B horizon may be a medium tan and the C horizon considerably lighter.

In some instances the soil color can be attributed to the presence of chemicals in sufficient quantity to stain the soil. Soils in much of the state of Georgia are red, because they contain iron oxides. In Arizona or New Mexico, the soils tend to be almost white, or a very light gray, because of the abundance of calcium and gypsum. In areas of high precipitation, the soils are light in color, because both chemicals and organic matter have been leached out.

The texture of a soil is determined by the size of the particles of which it is composed. Pedologists consider soil to range in grain size from the smallest and finest particles (clay, silt, loam) to the largest, granules of sand. If the particles are as large as gravel, the mixture is not true soil. Why is the texture of a soil important? It is because the texture influences the movement or retention of soil water. Sandy soils tend to lose moisture rapidly, and soils containing clay tend to resist absorption of water for some time (although in clay soils the air spaces eventually fill and the moisture will then remain). The best agricultural soils are somewhere in between, with a texture that allows both gravitational and capillary water to move freely through the horizons.

A soil's structure is determined by how the soil particles align themselves. If they form in thin, flat sheets or "plates," we say the soil's structure is **platy.** If the soil is composed of minute granules with rounded shapes, the soil is **granular.** Uneven, irregular, sharp-edged grains are called **blocky,** and if the grains are aligned in straight lines, taking on the many-sided shape of a prism, the soil is termed **prismatic.**

SOIL-FORMING PROCESSES   There are several soil-forming processes, but the three that are most important to humans are podzolization, laterization, and calcification. Each process operates differently, and each produces unique soils.

Podzolization   The process called **podzolization** (from the Russian *podzol,* meaning "soil under foot") occurs in cool, humid regions under a vegetative cover that is primarily composed of coniferous and broadleaf deciduous forests. Typically, the taiga forest regions of Canada and Siberia have podzolic soils. **Podzols** are acidic, with silica the prime mineral. Leaching removes most of the iron and aluminum oxides, and there is little humus in the A horizons. This results from the slow decomposition of the limited amount of plant and animal matter that does collect on the ground.

Soil Structure. (a) platy; (b) granular; (c) blocky; (d) prismatic.

There is little bacterial activity, owing to the cool climate, which lasts from eight to nine months each year. Because of the absence of humus and the presence of large amounts of silica, podzols are grayish to white in color.

Laterization The process of **laterization** (from the Latin *later,* "brick") produces soils called **latosols** (and, sometimes, *laterites*) and occurs in warm, humid climates, such as the humid subtropical and tropical rainforest areas. Leaching is common because of the heavy precipitation, resulting in the removal of the important mineral silica. The most widespread latosols are composed of iron and aluminum compounds (thus their name, *pedalfer,* from *ped,* "ground" or "soil," *al,* "aluminum," and *fer,* "ferrous"). Soil profiles in areas of laterization are very deep, and the A horizon may sometimes extend as much as 20 to 30 feet (6 to 9 meters) in depth. B horizons are difficult to distinguish as a separate zone. Groundwater movement is active because of constant replenishment from above.

Calcification The name **calcification** was derived from the dominant mineral combination in the process, calcium carbonate. Calcification occurs in climate regions that range from arid to semiarid to humid and are noted for their grass cover. The roots of the grasses reach far down into the B horizon and bring back calcium to the surface. Dying grass and good bacterial activity supply large quantities of humus, which becomes well mixed throughout the A horizon. The depth of these soils is not as great as the lateritic soils, but it is far more fertile and, therefore, more useful for agriculture and livestock ranching. Mixing carbon dioxide with the ever-present calcium produces calcium carbonate, which fosters lush grassland vegetation. The most famous of these areas are the fertile steppes of the Ukraine, the pampa of Argentina, and the Great Plains of the United States.

Soil Classification

Pedologists have developed a number of ways to classify soils beyond the processes we have just discussed. The classic system, which evolved out of research by Russian and American pedologists during the late nineteenth and early twentieth centuries, divided the earth's soils into three major groups or zones: zonal, intrazonal, and azonal. The latest system, called the 7th Approximation and destined to become the standard, divides soils into 10 basic groups. We shall consider both of these systems briefly, and as we do, it would be wise for you to consult the maps of vegetation (Figs. 7–2 through 7–4), climate (Figs. 8–2, 8–5, 8–9, 8–12, and 8–17), and land-forms (Fig. 10–3).

THE 7th APPROXIMATION The **7th Approximation** classification system was devised in 1960 by pedologists in the Soil Conservation Service of the U.S. Department of Agriculture. The 10 soil orders are based on the environmental controls that produced the 38 suborders (Fig. 12–2). Following are the 10 basic orders.

1. *Inceptisols* Soils with indistinct horizons resulting from a short evolutionary time, as in tundra areas, or those evolving from parent material that is resistant to change, as in mountainous and volcanic regions.

2. *Alfisols* Mixed acid–alkaline soils found in subtropical to tropical climate areas in the main agricultural regions of the world.

3. *Spodosols* Soils found in cool, humid climate regions; includes soils formed by podzolization.

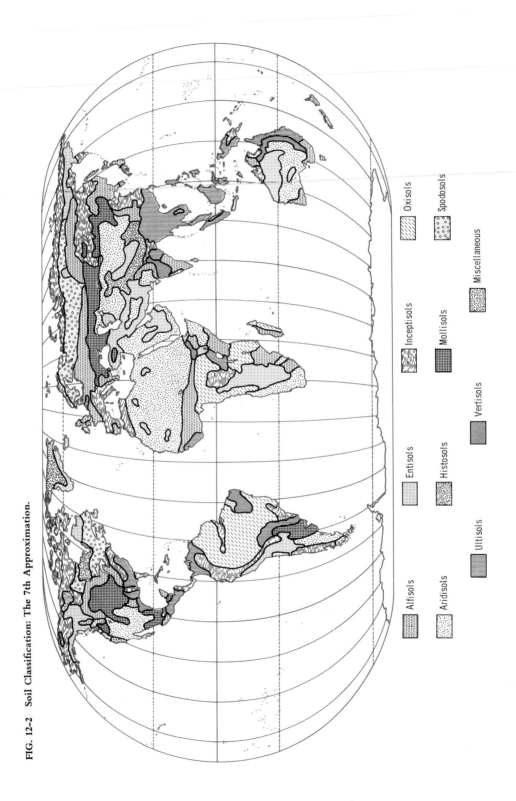

FIG. 12-2 Soil Classification: The 7th Approximation.

Alfisols

Aridisols

Entisols

Histosols

Inceptisols

Mollisols

Oxisols

Spodosols

Ultisols

Vertisols

Miscellaneous

4. *Utisols* Soils found in tropical wet and dry regions that are enriched with subsurface clay layers; considered to be good agricultural soils.

5. *Oxisols* Tropical rainforest soils found in lowland areas; primarily lateritic in composition.

6. *Mollisols* Grassland soils found in the prairie and steppe regions of the world, with covers of forest on the wetter margins and grasses on the drier areas; the chernozems are included in this order.

7. *Aridisols* Desert soils that are deficient in both moisture and organic matter, so usually not used for agriculture.

8. *Entisols* Includes new soils that have not had time to develop horizons, or old soils that are thin and stony and are found in mountain regions, interior deserts, and some interior permafrost areas.

9. *Vertisols* Clay soils with a brittle and cracked surface.

10. *Histosols* Soils that form in swamps or bogs.*

THE GENETIC SOIL CLASSIFICATION   The genetic soil classification system (Fig. 12–3) may be rather ancient, but it is relatively easy to understand and has served pedologists well. It is based on the stages of development of the world's many soils. Well-developed soils are called *zonal;* immature soils, *azonal;* and intermediate soils, *intrazonal.* As we consider these classifications, we shall also look at some of the more important soil groups within them.

Zonal soils   The most highly developed of all soils, as well as the most widespread, are **zonal soils.** They include both fertile and infertile soils, depending on the climate region in which they are found and the type of vegetative cover that dominates. Following are subclassifications of zonal soils.

1. *Forest soils* These soils are represented by podzols and latosols. The podzols range from light grayish soils that develop under the coniferous forests of the higher latitudes (the taiga forests of Canada and Siberia) and some mountainous regions (southern Chile). A more fertile, grayish-brown version evolves under broadleaf deciduous forests on the North European lowlands (including much of Great Britain, France, and Germany) as well as the Ohio Valley region of the United States. The higher fertility of this podzol results from a lower rate of leaching. A red-yellow podzol develops in warmer, humid areas of the midlatitudes (southeastern United States and southern Brazil). The latosols are tropical soils that are low in fertility, owing to excessive leaching and a low humus content. These reddish, reddish-brown, and yellow-brown soils are found in most tropical rainforest regions (Brazil, West Africa, Southeast Asia and the East Indies, Malagasy, the Caribbean, and portions of Mid-America).

*For a thorough explanation of this new system, see: U.S. Department of Agriculture, Soil Conservation Service (1960), *Soil Classification: A Comprehensive System: 7th Approximation* (Washington, D.C.: U.S. Government Printing Office); and, J. G. Cruickshank (1972), *Soil Geography* (New York: Halsted Press; Wiley).

FIG. 12-3 Soil Classification by the Genetic System.

Chernozems   Tundra   Desert

Latosols   Mountains   Prairie

Podzols   Sierozems   Alluvium

338

2. *Grassland soils* This group includes three important classes: the chernozem soils, the prairie soils, and the reddish prairie soils. The chernozems develop in warm-to-hot, semiarid climates and are covered by grasses that provide excellent fertility (as evidenced by the productivity of the northern Great Plains of the United States and by the recent increase in agricultural activity throughout the steppe regions of Russia). Areas bordering the steppe regions produce the prairie soils, which are also extremely fertile, and which can be used for growing grains or grazing livestock. The reddish prairie soils develop around the borders of the tropical and subtropical climate regions and are typical of the tropical wet and dry climates (the veldt areas of East and South Africa, eastern and southern Australia, Argentina's pampa, northwestern India, Cambodia, northern Algeria, and the southern portion of the Great Plains of the United States).

3. *Desert soils* Included in this group are the chestnut brown, gray desert, and red desert soils. Chestnut brown soils develop in sections of the steppes where the precipitation is light and the cover is short grasses; thus they are not considered to be very productive soils. Gray desert soils (also called *sierozem*) develop in cool, dry desert areas, such as Patagonia (Argentina), the Takla Makan and Gobi deserts of Tibet and Mongolia, the region east of the Caspian Sea, and parts of Wyoming and Colorado east of the Rockies. Red desert soils develop in hot, arid desert climates, such as in North Africa's Sahara, the Arabian Desert, the Thar Desert of India and Pakistan, Australia's Great Sandy, and the Sonoran of northern Mexico and the American Southwest.

4. *Tundra soils* These soils develop in the cold permafrost regions north of where trees will grow. They are found in Canada and Siberia in lands bordering the Arctic. The vegetative cover is composed of lichens, mosses, sedges, and dwarf willows, and the soils are not fully developed, except in a few places where the development of horizons is limited by the permanently frozen subsurface. In some areas there is humus, but it reaches a depth of only about 9 inches (22.5 centimeters). Because of the long winters, short summers, and low annual range of temperature, commercial agriculture is impractical.

Intrazonal soils Less developed than the zonal soils, **intrazonal soils** show signs of immaturity. The parent material and the terrain exert control over these soils, as is indicated by the three suborders:

1. *Hydromorphic soils* These are the waterlogged soils found in swamps and bogs, or they are found in meadows having better drainage, but also heavy clay deposits.

2. *Halomorphic soils* These soils are found in dry climates, especially where the rate of evaporation exceeds the precipitation. They include salty soils (solonchak) and alkaline soils (solonetz).

3. *Calcimorphic soils* These soils have a high lime content (calcium carbonate) and, because of their grassy covers, a high humus content. They include brown forest soils and rendzinas.

Typical Azonal Soil. Fertile alluvium deposited by a river over an original floodplain constitutes typical azonal soil.

Azonal soils The most immature (least developed) of the soils are the **azonal soils**; but they are not, however, necessarily the poorest in fertility. Although they may not have had sufficient time to develop definite horizons, some—alluvial soils—may be very fertile if their source region was high in mineral or organic content. Azonal soils that are found on slopes are thin and stony and are called *lithosols*. Those that are sandy (as in deserts) are called *regosols* (from the Arabic word *reg,* "sandy desert").

Review and Discussion

1. What is soil?

2. Why is climate considered the most important soil-forming control?

3. Explain the difference between gravitational and capillary water movement.

4. Explain the terms: *caliche, leaching, hardpan.*

5. What is humus?

6. Why is the parent material important in soil evolution?

7. Why is time important in soil development?

8. Why is alluvium sometimes considered fertile soil?

9. How is terrain important in the formation of soils?

10. What is a soil profile? Why is a consideration of soil horizons important?

11. Which horizon contains the most nutrients?

12. Explain the differences between these terms: *zone of eluviation* and *zone of illuviation.*

13. Why do soils vary in color?

14. Explain the difference between these soil structure types: platy, granular, blocky, prismatic.

15. Explain these soil-forming processes: podzolization, laterization, calcification.

16. Based on the brief explanations of the two major soil classification systems in use today, which do you prefer? Why?

17. What is the difference between podzols and latosols?

18. Why do soils develop differently under coniferous forests and grasslands?

19. Explain the difference between these terms: *hydromorphic, halomorphic, calcimorphic.*

20. Why are some azonal soils important to mankind?

Additional Reading

Baldwin, Mark, Charles Kellogg, and James Thorp; U.S. Department of Agriculture (1938). *Soils and Men: Yearbook of Agriculture, 1938.* Washington, D.C.: U.S. Government Printing Office.

Bunting, B. T. (1965). *The Geography of Soil.* Chicago, Ill.: Aldine.

Cruickshank, J. G. (1972). *Soil Geography.* New York: Halsted Press; Wiley.

Kellogg, Charles (July 1950). "Soil," *Scientific American* (offprint no. 821).

Miller, C. E., L. M. Turk, and H. D. Foth (1959). *Fundamentals of Soil Science.* New York: Wiley.

U.S. Department of Agriculture (1957). *Soil: The Yearbook of Agriculture, 1957.* Washington, D.C.: U.S. Government Printing Office.

———, Soil Conservation Service (1960). *Soil Classification: A Comprehensive System: 7th Approximation.* Washington, D.C.: U.S. Government Printing Office.

13

Embossed by Running Water

And see the rivers how they run
Through woods and meads, in shade and sun,
Sometimes swift, sometimes slow,
Wave succeeding wave, they go
A various journey to the deep,
Like human life to endless sleep!

GRONGAR HILL, JOHN DYER

The term *running water* can mean different things to different people: a dripping
faucet, a flowing stream, or the movement of groundwater deep beneath the earth's
surface. It can mean urgently needed moisture for food crops and natural vegetation,
and it can mean life for humans. Hawaiians of bygone years kneeled along the banks
of a stream and offered this prayer of thanksgiving:

> *One question I ask of you:*
> *Where flows the water of Ka-ne?*
> *Deep in the ground,*
> *In the rushing spring,*
> *In the ducts of Ka-ne and Loa,*
> *A wellspring of water to quaff!*
> *A water of magic power—*
> *The water of life!*
> *Life! O give us life!*[1]

The running waters of the earth, whether on the surface or underground, do
more than supply us with the "water of life." They have efficiently embossed the
surface and etched the deeper layers to provide us with magnificent landscapes of
rugged highlands and fertile valleys. Running water is only one of the earth's
sculpturing agents, but it is the most active.

Running waters are to be found everywhere—in every physiographic region and
in all latitudes—and they are persistent in their endless quest to level the land. It
would appear that their function is to counter the tectonic forces that operate to
raise and wrinkle the earth's surface.

How does all this happen? Why are streams so active? What kinds of activities
do they perform? How can we recognize the surface features created by running
water and know immediately that they are not the result of the other erosional
forces, such as blowing wind, grinding glaciers, or the heaving waters of the oceans?
This chapter will provide some of the answers.

Sculpturing Processes of Running Water

The sculpturing processes of running water—erosion, transportation, deposition—are
important by-products of the hydrologic cycle. As part of the endless process, the
water moves through the atmosphere from water bodies to the land, is precipitated,
either soaks into the groundwater systems or runs off the surface toward some distant

lake or ocean, and eventually returns to the atmosphere. As water moves along the earth's surface, it accomplishes a considerable share of the gradational activities of erosion.

EROSION   There are three methods of **erosion** (from the Latin *erodere,* "to gnaw out") by running water: splash erosion, sheetwash, and stream erosion.

Splash erosion is created by the impact of millions upon millions of raindrops falling on the earth's surface. As raindrops strike unprotected soil particles, the particles are dislodged from their original position and, sometimes, are thrown into the air. When they fall back to the ground, they are easily picked up by running water and transported elsewhere. Because water-retaining silts and clays, as well as bits of humus, are removed, the structure of the upper soil layer is altered. The amount of precipitation, the porosity of the surface layer, and the gradient (degree of slope) have a lot to do with the effectiveness of splash erosion. Obviously, soft and weakly structured rock layers are broken up more rapidly than is resistant rock such as granite, yet all are modified in time. If a sturdy vegetation cover (or a mulch applied by humans) is present, splash erosion will be delayed for long periods and, sometimes, actually prevented, thus saving agricultural soils for further use.

Sheetwash occurs after sufficient water has fallen to the ground, collecting in pools or a thin sheet, and responding to the pull of gravity, and begins to flow down a slope. The type of rock, the cover of vegetation, and the steepness of the gradient all influence the ability of sheetwash to perform erosional activity. If the soils are composed of fine, easily transported sediments, erosion will commence almost immediately. If the gradient is steep, gullying can transform a once-productive farming region into a wasteland. Farmers all over the world have tried to prevent gullying by using methods such as contour plowing and planting crops that tend to diminish the power and destructiveness of overland flow.

Stream erosion is accomplished by three processes, two of which (hydraulic action and corrasion) are physical; the other, corrosion, is chemical. As water flows down through its channel (streambed), it picks up small grains and pulls gravels and pebbles along with it, thus downcutting (eroding) the streambed by **hydraulic action.** As the grains, gravels, and pebbles are flung against the streambed, its banks and any large rocks in the way are reduced in size by the abrasive action of **corrasion** (from the Latin *corradere,* "to scrape together"). When the rushing waters of a stream pass over a hardrock base, then abruptly move across a section of soft, loose materials, a pothole is dug out, and the flowing water becomes turbulent. **Corrosion** (chemical decomposition) is more apt to occur during movement of groundwater, especially in limestone areas and where carbonic acid is prevalent, or in streambeds where the surface flow is intermittent and some of the chemical weathering processes can operate.

The velocity of a stream depends on three factors: steepness of gradient (slope), volume of water, and the load. In a sense, these factors are closely tied together, like the links of a chain. If any one is altered, the entire chain is altered. A slow-

moving stream, one that travels at less than 0.5 mile (0.8 kilometer) per hour, can
carry only a light load of fine silts and sands, but if heavy rainfall upstream were
to increase the water's speed to a velocity of 4 or 5 miles (5 to 8 kilometers) per
hour, the stream could carry rocks with a diameter of about 3 to 6 inches (8 to
13 centimeters). If the velocity were increased to a speed of over 10 miles (16
kilometers) per hour, large boulders could be rolled along the streambed. Sometimes
a stream erodes its bed down to such a gentle slope that the stream can no longer
downcut and is said to have reached baselevel.

Baselevel (Fig. 13–1) is the lowest level to which a stream or river can lower

itself: a plain at the mouth of a canyon, another stream or river, a lake, or the ocean.
Since a stream cannot flow uphill, it can only degrade its bed until it joins the other
body of water and its downward erosional power is ended. The time required for
this process varies with the size of the stream. A heavy downpour over a weak rock
area that has a steep slope will create what might almost be termed "instant valleys,"
deep, steep-sided gullies, and by storm's end, the stream occupying the gully will
have reached baselevel. On the other hand, a stream that must travel a great distance
over gently sloping terrain will have to struggle for a long period of time to attain
baselevel. Once baselevel is reached, for example, at the mouth of a stream valley,
the stream will cease degrading and begin to reduce its entire course to baselevel.

The ultimate baselevel could be lowered only if the entire region were uplifted
or if the lake, river, plain, or ocean into which the stream flowed were to have
its surface level reduced. If either one of these things happened, the stream would
immediately begin to reassert itself, degrading would commence again, and a new
baselevel would ultimately be established. What if the terminal point were raised?
If the surface of the river, lake, plain, or ocean should rise, the stream immediately

FIG. 13–1 Baselevel.

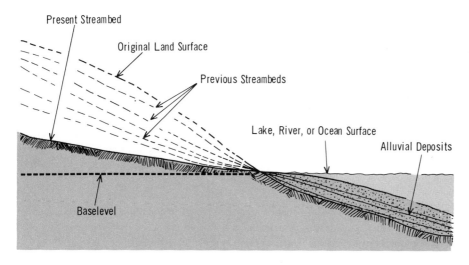

would be forced to begin to fill in the old and lower bed with alluvial deposits and, so, establish a new and higher baselevel. The Colorado River has had its baselevel changed many times in its journey through the Grand Canyon, and the evidence is there for all to see: layers of marine sediments (left by oceans that once covered that area) resting atop shales deposited by river waters, which rest atop other marine sediments.

Graded streams are those that have attained a degree of balance with the important erosional factors of volume, slope, and load. A graded stream no longer degrades its bed vertically, and its erosional activity is confined to lateral cutting of the sides of the valley. When the amount of water supplied combines with the slope just barely to transport the materials that have been eroded laterally, the stream is said to have reached *equilibrium*. This means that it has become a graded stream. However, if any of the three factors should be changed, the stream would be thrown off balance and would no longer be in a state of equilibrium. This could happen if a stream that had been eroding weak rock materials worked its way down to hard, resistant bedrock and thereby could move without being forced to carry a load. It could also happen if there were a climate change, with precipitation increasing the stream's volume of water, or if there were an increase in the stream's slope owing to an uplift of the land.

EROSION LANDFORMS Interesting examples of the results of the erosional power of streams include waterfalls, incised meanders, terraces, and stream valleys. The last one, stream valleys, is, perhaps, the most important, and we shall consider this in more detail when we discuss the relationship of the geographical (or geomorphic) cycle to the development of stream valleys.

Waterfalls can result from a stream's passage over an escarpment, such as a recently uplifted surface, or they can be the result of the process of continued erosion of soft, loosely compacted beds of shale or sandstone that join along a boundary line with resistant rock (sometimes called "cap rock"). The escarpment or cliff is continually undercut, resulting in the fall of the overhanging portions and the subsequent receding of the cliff face. Niagara Falls is an excellent example of a "retreating" waterfall.

Incised meanders (Fig. 13–2) are the result of a gradual uplift of a region, thus forcing the stream to continue a vigorous downcutting in its previously determined meandering (wandering from side to side) pattern.

Stream-cut terraces are the result of a stream's wandering across its valley floor, eroding the sides as it moves. If the land were to be upraised, the stream would resume a vigorous downcutting and so create a new, but narrower, valley floor. If this process continues, a series of the old valley floors will appear much like shelves fastened to the valley walls.

STREAM TRANSPORT After rock materials have been weathered and eroded, the stream must then transport the material to another place. The finest particles are carried in suspension, and the heavier grains or fragments are bounced or rolled along. The speed of the stream determines the amount of the load and what kind of material

FIG. 13-2 Incised Meander of the Colorado River in the Grand Canyon.

is carried, but all streams transport something. Even those seeming to be clear carry chemicals dissolved from the rocks. When you think of stream transport, you probably picture rivers such as the Mississippi, Amazon, or Nile, because they carry tremendous loads of soil eroded from faraway highlands and seem to move sluggishly. Jet pilots flying from Africa to South America along the equator say that you can see the continent of South America long before you reach it; that is, you see the reddish colored soil-laden outflow from the Amazon River some 100 miles (160 kilometers) before you can see Brazil.

STREAM DEPOSITION   We have been discussing the gradational (wearing down) force of erosion. Now let us consider the opposite force, deposition, which also works to level the land. Deposition does not tear or wear down, however, for this process is one of **aggradation,** a building up. One might say that erosion is destructive; deposition is constructive. Fluvial (from the Latin, *fluvialis,* "river") hydrologists tell us that sediments will be deposited out of a stream of moving water when the velocity of the stream is slowed to where it can no longer continue to transport material. This can happen when a stream flows out of a canyon onto a flat or a nearly level plain or as it wanders back and forth (meanders) in its valley, eroding one side and depositing on the other. A number of interesting landforms are created by this laying down of **alluvium** (stream-deposited sediments).

Alluvial fans (Fig. 13–3) occur when a stream rushes out of a steep canyon, and its velocity is slowed as it strikes a gently sloping area, forcing it to deposit its materials. The first materials to be sedimented are boulders, then smaller rocks, gravels, sands, and, finally, silt and clay particles. The initial shape is that of a cone, but as the stream continually blocks its own path with its deposits, it must seek new courses, and the larger, fan shape results. If there are a number of canyons fronting a broad valley or plain, the fans developing at the mouth of each canyon may coalesce—that is, come together—and the combination may stretch along the foot of the mountains for many miles. This alluvial landform is termed a **piedmont alluvial plain.**

Deltas form when a stream or river flows into a large lake or ocean, and its movement forward is immediately slowed (Fig. 13–4). The resulting buildup of deposits is almost identical to those of alluvial fans. Sometimes deltas form within the mouths of rivers, perhaps as part of a submerged coastline where the ocean has intruded into the river's valley. This delta is termed an **estuarine** (from the Latin *aestuarium,* "where the tide meets the coast"), because the place of meeting is at the mouth of a river, an estuary. In time, a simple delta (Δ) shape may form as the sediments are pushed out into the lake or sea, but its future shape will be regulated by a number of factors. If the incoming stream is confronted by swift offshore currents, an **arcuate** delta will form; if the currents do not have a consistent direction,

FIG. 13-3   Alluvial Fans. (on page 351) (*a*) alluvial cone; (*b*) alluvial fan; (*c*) piedmont alluvial plain; (*d*) cross section of a typical alluvial fan.

(a)

(b)

(c)

Mountain Slope

Streambed

Stream-Deposited Alluvium

Floor of Valley or Plain at Foot of Mountain

(d)

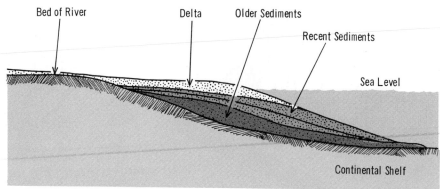

FIG. 13-4 Formation of Deltas.

but are constantly changing, a **cuspate** delta will come into being. If the waters offshore are calm and do not have the power to disturb the even flow of sediments, a **bird-foot** delta takes shape.

Some of these deltas are quite massive, extending many miles out to sea, complete with numerous *distributaries* (streams that branch off the main course). Some have heavy concentrations of population. A good example of the importance of deltas in the history of civilization is the Nile delta. The fertile soils of this area drew people to it, and the delta became Egypt's "cradle of civilization."

Floodplains (Fig. 13–5) are the result of a considerable widening of a valley by a meanering river. As the river flows across the land, it meanders, cutting away at the banks on one side and depositing sediments on the opposite side. In Fig. 13–5 the points of heaviest erosion occur at points A, C, and E, because the swiftest part of a meandering stream or river is on its outside edge. Deposition occurs at points B, D, and F, for the inside portion of a river is the slowest, often forming swirls of backwater. Because this type of stream erodes laterally, there is little vertical, or downward, cutting, except during times of floodstage (when the volume of the river is tremendously increased). Generally, floodplains evolve through stages of development, widening through time and forming features such as natural levees and alluvial terraces.

When floodstage arrives, the height of the waters is increased by waters swollen by rain, which may have fallen in a distant area, until the river's banks can no longer contain the increase. Waters pour over the banks, carrying their load of sediments onto the valley floor on both sides of the river. As the speed of the rushing waters is checked, the heavier particles are deposited, then the lighter and finer ones. Thus a mound paralleling the river is built up. This **natural levee** will contain the stream or river until the next high floodstage. People combat this problem by building artificial levees (sandbags or piles of boulders) to the top of the natural levees. Despite the dangers of residing in floodplain areas—the seasonal threat of flooding and the resulting loss of farm crops and human lives—these areas of the world appear to have the greatest concentrations of human population.

Bird-Foot Delta of the Tongariro River, New Zealand.

The Broad Delta of the Nile River.

Arcuate Delta, Chinitna Bay, Alaska.

Estuarine Delta. The delta has formed at the mouth of the Colorado River where it empties into the Gulf of California.

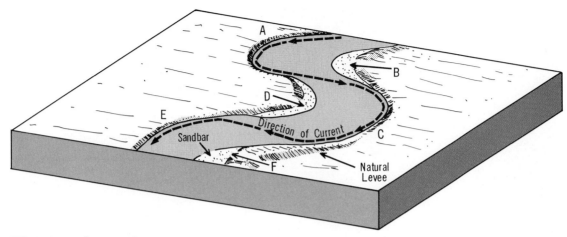

FIG. 13-5 Development of a Floodplain.

Alluvial terraces are composed of alluvium deposited by streams, arranged like the stream-cut terraces along the sides of stream valleys. The main difference between these two different types of terraces is that the stream-cut terraces are usually "matched"; that is, on each side of the valley, the terraces are the same age, so they have the same elevation. Alluvial terraces, on the other hand, are not normally matched, because the deposits on each side of the valley are laid down at different times and under different conditions of sediment load and water volume. As the land is uplifted or as the stream downcuts, the terraces are exposed.

Oxbow lakes (Fig. 13–6) are to be found in relatively old floodplains that have had enough alluvium deposited through the years to raise the general level of the valley floor. The river no longer does much downcutting as it meanders sluggishly through the valley. When one of the meandering loops is bypassed, or cut off from the main stream flow, an oxbow lake (so called because its shape is similar to yokes worn by oxen pulling wagons) is formed. If enough water continues to seep into the lake, it will remain a lake; if not, the oxbow will slowly be transformed into a marshland and, eventually, a grassy meadow or a farmer's field.

Development of Stream Valleys

Valleys can be found almost everywhere, but they will not all have the same shape and size because of a number of controlling factors, such as the composition of the rock layers over which the stream must pass, the cover of vegetation that may hinder the movement of the running waters, the gradient of the slope, the regularity and amount of precipitation, the rate of evaporation, and time.

STREAM-VALLEY EVOLUTION There has been considerable discussion during the twentieth century over the question of the evolution of stream valleys. Near the

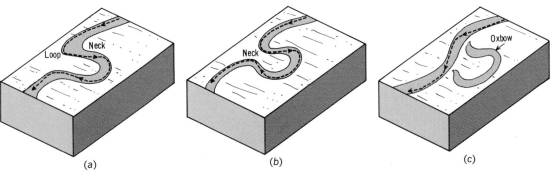

◄---- Direction of current

FIG. 13-6 Development of an Oxbow Lake. (a) A meandering river carves great loops on the valley floor; (b) the meander loops close in on each other; (c) the river cuts through the narrow neck, and the meander is cut off, becoming a cutoff meander, or an oxbow lake.

end of the nineteenth century, a geologist-turned-geomorphologist, William Morris Davis, formulated a theory to explain the leveling of the earth's surface, a process he termed *the geographical cycle.**

After a long period of field study, Davis came to the conclusion that after the initial uplift of a landmass, running water would immediately go to work to bring it all down to a final, flat plain, the **peneplain** (derived from another of his terms, *penultimate plain,* meaning "almost to the end"). He believed that someday all the mountains, highlands, plateaus, and hills would be gone, and only a flat plain would be left. Davis did suggest that a few rocky remnants might be left jutting up from the peneplain, resistant rock mounds he called *monadnocks* (after New Hampshire's Mount Monadnock). However, his general theory of the ultimate end of the cycle of erosion was avidly opposed. Earth scientists of today do not accept his theory, because they say that the earth's surface does not reach a standstill. Diastrophism, vulcanism, and plate tectonics give ready proof, they claim, that the landmasses are continually rising at the same time they are being graded down. They say that individual streams may achieve baselevel, but only for a time, and broad areas will experience infinite crustal movements, thus preventing the attainment of a peneplain.

This does not mean, however, that we cannot apply Davis's concept of a geographical cycle to the evolution of stream valleys. Some valleys show their youthfulness by their rugged shapes, and others show their age by the flatness of their landscapes. The uplifting of landmasses and the subsequent grading down is a continuing process. Valleys, like mountains, work their way through a cycle of birth, life, and death, from youth, to maturity, to old age. After death (the peneplain) comes rebirth, and the wheel revolves.

Streams coursing down the slopes of young mountains with high elevations and steep gradients are strong and vigorous. The channels of these **youthful streams** are fairly straight, and their valleys are narrow and V-shaped. The land between two stream valleys (called *interfluves,* "between the fluves" or "between the streams") is high and broad. The general relief of the area is sharp and bold, not yet rounded by erosional forces. There is little in the way of lateral erosion, because the steep slopes discourage all but vertical cutting. As time passes, the *drainage area* (the valley between two ridges that catches the various forms of precipitation) will supply additional water in the form of tributaries that add to the widening of the valley.

If there are numerous obstacles in the stream's path, such as granitic dikes or other resistant rock masses, the valley may contain many waterfalls and lakes.

A **mature stream** is no longer vigorous in its vertical attack upon the valley floor; it has arrived at the stage of equilibrium. The valley, rather than being steep sided, would have been cut laterally by the stream for sufficient time to narrow the interfluves and widen the drainage area. It would be broad and flat, forming a well-developed floodplain with, perhaps, some small lakes, natural levees, and sand-

*William Morris Davis (November 1899), "The Geographical Cycle," *The Geographical Journal,* 14, no. 5, 481–504; available in Bobbs-Merrill reprint no. G-48. For opposing views, see: Kirk Bryan (December 1940), "The Retreat of Slopes," *Annals, Association of American Geographers,* 30, no. 4, 254–268; Bobbs-Merrill reprint no. G-28.

Youthful Stream Valley of the Neretva River, Yugoslavia.

bars. In such a valley, tributaries are common, and the drainage system is conducive to the creation of a meandering stream. There may even be a few oxbow lakes.

By the time a stream valley has reached the stage of old age, it has done just about all it can, unless some form of rejuvenation occurs. An **old-age stream** usually moves slowly, almost sluggishly, as it swings in great loops from one side of its wide floodplain to the other. It may be wide and shallow, smoothly functioning in a state of equilibrium, except during times of high water (floodstage) when it overflows its banks and natural levees to inundate the land beyond. Interfluves will be low in elevation and in local relief, smoothly rounded, narrow, and widespaced. There will be marshes, oxbow lakes, stream-cut and alluvial terraces, numerous tributaries, but a poorly developed drainage system, excellent farm land (if artificial levees can contain the seasonal rampages of the river), and a broad delta at the river's mouth.

Earth movements can cause disruptions at any point in the evolution of a stream valley. If there has been diastrophic uplift, as has occurred along the coast of Norway, an old-age stream will be rejuvenated and will begin carving out a youthful valley. If there has been a general subsidence of a large area, such as happened in the

Mature Stream Valley of the Fremont River, Utah.

Old-Age Stream Valley near Alsiak Bay, Alaska.

Mississippi River region, the floodplains will be widened, and the river will flow even more sluggishly. In this case, a youthful stream valley might move directly into an old-age situation, bypassing the mature stage.

CLIMATE AND STREAM-VALLEY DEVELOPMENT Stream valleys in humid (moderate-to-heavy precipitation) climate regions are quite different from those in dry or semiarid regions, primarily owing to the vegetative cover and the amount and constancy of stream flow. Stream valleys in dry climates show the effect of thunderstorms: Sudden torrents are unleashed, attacking the looseknit soils that, in such areas, have no vegetation to hold back the surging waters. Youthful valleys in humid areas are not as rugged as those of dry areas, and old-age valleys of dry areas are filled with alluvial deposits that seem to rise to the mountain tops. Humid areas are noted for their hills and low, rounded mountains, while dry areas are noted for their deeply dissected plateaus and mesas.

Groundwater Etchings

During our earlier discussion of the earth's waters, we learned that groundwater supplies mankind with fresh water for drinking and for agriculture. But is that all we can say about groundwater? Does it serve mankind in other ways? The answer to this question is in the affirmative, because groundwater performs in ways similar to surface water: It furnishes salts to the oceans and creates interesting and sometimes fantastic erosional and depositional landforms.

EROSION BY GROUNDWATER When there is precipitation over a land composed of massive beds of limestone, water entering the ground is transformed into a powerful erosional force. Carbon dioxide in the air unites with the water to form carbonic acid, which literally eats its way through the limestone as it dissolves the calcium carbonate and carries it away in solution. Figure 13–7 shows the typical landforms created by groundwater that result in a landscape called **karst** (after the region of that name in Yugoslavia where the work of groundwater on limestone beds was first seriously studied). This, then, is another cycle of landform evolution. The stage of youth is shown in Fig. 13–7(a) and (b), where numerous surface streams sink into the ground and begin the work of dissolving the limestone. As the surface layers are attacked and dissolved, a number of depressions, termed **sinkholes,** are formed. As time passes, some of these sinkholes increase in size and become deeper. These larger depressions are called **dolines** (Fig. 13–7[c]). When further time passes and erosion continues, the dolines become enlarged as surface streams run over their sides and disappear from view as they flow through underground channels to join the water table.

The stage of maturity is reached when the karst area is dotted with large depressions, those resulting from the collapse of the roofs or ceilings of underground caverns. These steep-sided sinkholes (Fig. 13–7[d])—called *ponors* by the Yugoslavs, *swallow*

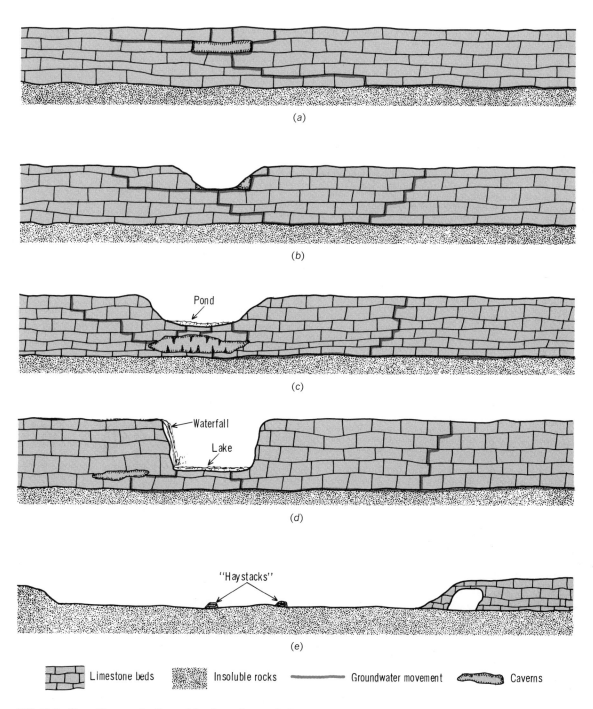

(a)

(b)

Pond

(c)

Waterfall

Lake

(d)

"Haystacks"

(e)

Limestone beds   Insoluble rocks   Groundwater movement   Caverns

FIG. 13-7 Karst Topography Created by Groundwater Action.

holes by the English, and *cenotes* by the Spanish—have flat bottoms that may contain marshes, ponds, or lakes.

By the time the region has reached what might be termed old age, the sinkholes will have coalesced, and the level of the land will be considerably lower than it was during its youth. Streams will once more flow across the insoluble rock surface that is now exposed, and the landscape of this *uvala* (or *polje*) will be dotted with the honeycombed remnants of limestone called *magotes* (or "haystacks"). There may even be a few natural bridges around the edges of the great valley (Fig. 13–7[*e*]).

DEPOSITION BY GROUNDWATER The most interesting depositional work done by groundwater can be seen only in underground caverns, such as those in the Mammoth Caves of Kentucky and Carlsbad Caverns of New Mexico. Here, you will discover a fantasy land of wierd and beautiful forms, created as the dissolved limestone from the overlying beds drips from the ceilings of the caves. As the moisture trickles down, deposits of calcium carbonate create fingerlike projections that hang from the ceilings (*stalactites*) and jut up from the floors (*stalagmites*). You may be interested to learn that one of our common building materials for bathroom walls and floors, travertine, comes from limestone caverns like these.

Limestone Formations. Stalactites, stalagmites, and draperies have formed in the Frozen Niagara section of Mammoth Caves, Kentucky, as a result of the slow percolation of water through the cracks and joints of the overlying limestone.

Review and Discussion

1. Explain the difference between these terms: *corrasion* and *corrosion.*

2. What is splash erosion? What is sheetwash?

3. What three factors determine the velocity of a stream?

4. What is the difference between a stream-cut terrace and an alluvial terrace?

5. Define these terms: *aggradation, gradation.*

6. Explain the structure and development of an alluvial fan and a delta.

7. What is a piedmont alluvial plain?

8. Describe the development of a floodplain.

9. Explain these terms: *natural levee, oxbow, meander.*

10. Explain baselevel.

11. What is a graded stream?

12. What is a peneplain? A monadnock?

13. What is an interfluve?

14. What is the effect of climate on stream-valley development?

15. Explain the difference between dolines, ponors, and uvalas.

Note

[1] N. B. Emerson (1909), *Unwritten Literature of Hawaii: The Sacred Songs of the Hula* (Washington, D.C.: U.S. Government Printing Office).

Additional Reading

Ellison, W. D. (November 1948). "Erosion by Raindrop," *Scientific American* (offprint no. 817).

King, Lester C. (July 1963). "Canons of Landscape Evolution," *Bulletin of the Geological Society of America,* 64, 721–751. Also available as Bobbs-Merrill reprint no. G-116.

Mackin, J. Hoover (May 1948). "Concept of the Graded River," *Bulletin of the Geological Society of America,* 59, 463–512.

Sweeting, Marjorie (1973). *Karst Landforms.* New York: Columbia University Press.

Thornbury, William D. (1969). *Principles of Geomorphology.* 2d ed. New York: Wiley.

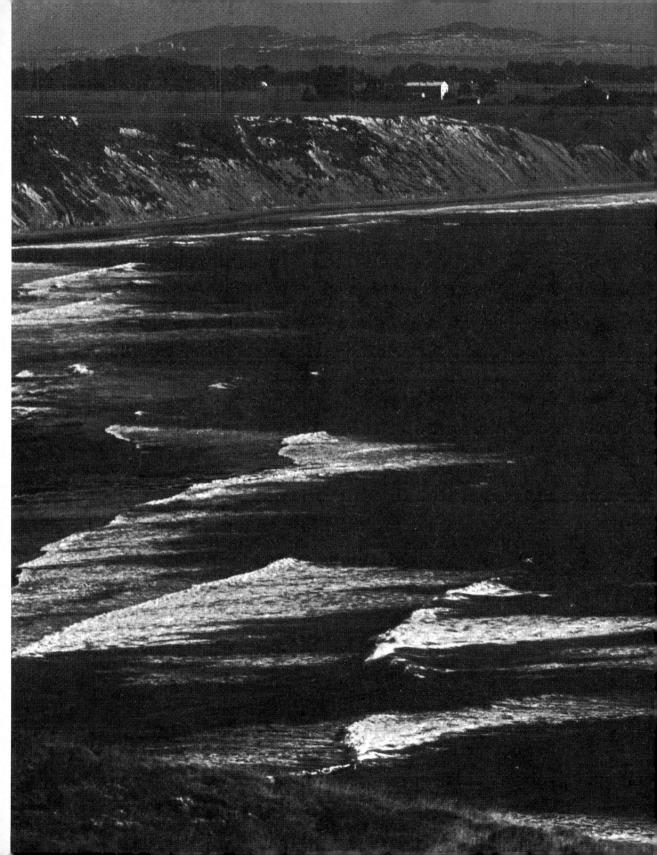

14

They that go down to the sea in ships, that do business in great waters; These see the works of the Lord, and his wonders of the deep. For he commandeth and raiseth the stormy wind, which lifteth up the waves thereof. They mount up to the heaven, they go down again to the depths: their soul is melted because of trouble. They reel to and fro, and stagger like a drunken man, and are at their wit's end.

PSALMS 107:23–27

Throughout the long history of man, many individuals have tried to fathom the mysteries of the "great waters," but it was not to understand the reasons behind the formation of the seas, the influence of the waters on their climates, or the sculpturing actions of the oceans on the lands. Because life was a daily struggle for survival for early peoples, the oceans were important as places that could be "farmed" with net and hook or used as "sea highways" to be traveled. While it was necessary for them to know of the rise and fall of tides and the movement of ocean currents, their knowledge and technology would not let them pursue the matter further.

The passage of time brought great changes. The study of the oceans was elevated to a respected place among the sciences by the work of such men as Benjamin Franklin, who charted the route of the North Atlantic's Gulf Stream; Charles Darwin, who traveled the oceans and, in 1840, wrote the epochal *Voyage of the Beagle;* and the American naval officer Matthew Fontaine Maury, who, in 1855, wrote the first oceanography text, *The Physical Geography of the Sea.* Modern oceanography became even more securely established when England's Royal Geographical Society sent the H. M. S. *Challenger* on a record-setting $3\frac{1}{2}$-year voyage to conduct oceanographic research in the Atlantic, Indian, Antarctic, and Pacific oceans. The voyage ended in May 1876, the ship having journeyed almost 69,000 nautical miles.[1]

A century has passed since that memorable event, and oceanographic research is continuing, as the nations of the world try to reach agreement on how they are to "use" the ocean waters and the sea bottom. It is important for us to understand what has been learned about these vital regions, so that we can keep abreast of developments as news flows from the international conferences. We shall begin expanding our knowledge of the oceans by examining that part of the ocean floor that is most accessible to us: the rim of the continental landmasses, the **continental shelf**.

Continental Shelves

The term *shelf* may be defined as a "flat ledge or horizontal surface usually fastened to a wall to serve as a support for objects of value." The definition is apt, because the continental shelves do support objects of value to mankind, such as rich petroleum deposits and many varied forms of sea life. The continental shelves, however, are not really affixed to the landmasses. They are a part of the land, an extension of the continents, forming a relatively flat but gently sloping margin of varying width. Figure 14–1 shows a typical cross section of the rim of the continents.

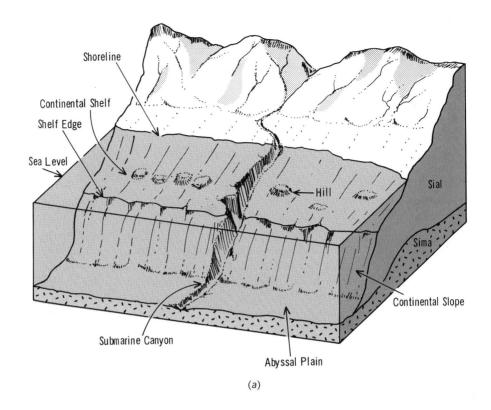

Shoreline

Continental Shelf

Shelf Edge

Sea Level

Hill

Sial

Sima

Continental Slope

Submarine Canyon

Abyssal Plain

(a)

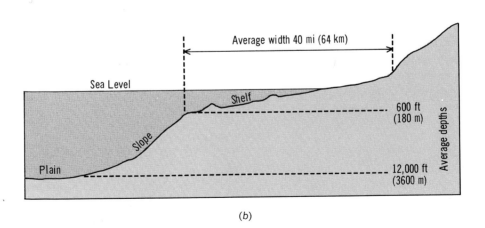

Average width 40 mi (64 km)

Sea Level

Shelf

600 ft
(180 m)

Slope

Average depths

Plain

12,000 ft
(3600 m)

(b)

FIG. 14–1 Rim of the Continents. (a) is a perspective view; (b) is a cross section. (Vertical scale greatly exaggerated.)

The term **shelf edge** marks the outer limit of the continental shelf and the beginning of the downward slope to the sea bottom beyond. This occurs, on the average, at a depth of about 600 feet (180 meters). This zone, from the surf line to the shelf edge, varies greatly in width, as you can see by studying a map detailing the world's continental shelves (Fig. 14–2). All countries lying in contact with the ocean do not have wide shelves. The width is near zero along some portions of the west coast of South America, but observe how wide they are around Great Britain and Ireland, northern Australia, and the East Indies. Some ocean-fronted countries have little or no sea life (fish, seaweed) or minerals to harvest, while others have shelves extending to 900 miles (1500 kilometers) that provide them with opportunities for bountiful harvests. Oceanographers estimate the average width of the world's continental shelves to be about 40 miles (64 kilometers).

ORIGIN OF THE CONTINENTAL SHELVES How did the continental shelves come into being? Geologists originally believed that they were the result of wave action attacking the granitic (sial) margins, wearing them down in some places, building them up in other areas. However, close examination of the structure of the shelves has shown that while the underlying beds are composed of granitic rocks, much of the covering material is made up of sedimentary rocks. This tells us that at some time in the past, perhaps many times, the level of the oceans was far below what it is today. Earth materials were carried down and deposited on the exposed shelves by streams, winds, and glaciers.

Geologists have proven that the earth experienced at least four great ice ages during relatively recent geologic time (about 2.5 million years). Within each ice age, tremendous masses of ice formed at the poles and spread toward the equator. In the Northern Hemisphere, the continental glaciers covered much of North America and Asia. From where did the water that formed the glaciers come? It is probable that moisture rising from the oceans was precipitated as snow over the polar regions, thus lowering the level of the seas. It has been estimated that sea level was from 200 to 500 feet (60 to 150 meters) lower than its previous level during these periods. When the climate changed and the ice sheets melted, water was returned to the oceans, and the level returned to normal.

It is apparent that the continental shelves of the world were alternately exposed and submerged, with each interval lasting for many thousands of years.*

DEVELOPMENT AND STRUCTURE OF THE CONTINENTAL SHELVES The continental shelves are quite complex in their structure. Not all of them meet the same specifications. Some have gravels and coarse sand in areas close to the shoreline, and fine clays and muds deposited farther out, near the edge of the drop-off point, the slope. Other have patches of gravels and clays arrayed at random across the width of the shelf. Why is there such a variation? There are a number of factors responsible,

---

* For one of the best discourses on the subject of oceanography see: Francis P. Shepard (1973), *The Earth Beneath the Sea,* rev. ed. (New York: Atheneum).

FIG. 14-2   Continental Shelves of the World.

such as the presence or absence of large streams or rivers, the type of rock on the land behind the shelf, whether glaciation occurred on or near the shelf, and the form and strength of the ocean currents operating offshore.

Rivers have been instrumental in building some stretches of continental shelves. As we have seen, large rivers can build extensive deltas at their mouths during the periods when the shoreline is submerged. When the sea level is lowered during an ice age, the rivers supply sediments to construct alluvial fans at their mouths. The shelves at the mouths of the Mississippi and Amazon rivers are much wider than shelves elsewhere along those coasts and are composed of fine soil particles, such as clay and some sand.

Glaciation in the vicinity, another force in the development of continental shelves, is evidenced by terrain that is not as flat or smooth as the average shelf. If a continental glacier actually moved across the shelf, it would leave its mark in the form of deep gouges and old river valleys that were scoured out by streams issuing from the glacier. The shoreline would look much like the rugged land surface, with many indentations engraved by the moving glacier. One would be likely to find sands, gravels, and rocks in the upper layer of the shelf (except where these deposits were covered by more recent deposition).

Topography can also be important, especially as a source from which earth materials are eroded and transported out onto the shelf. If the region behind the shelf is mountainous, the shelf might become dotted with rocky hills and its composition would be as varied as the composition of the mountains would allow. *Cobbles* (large, round, smooth stones), gravels, and coarse sand would be in plentiful supply.

Ocean currents can be very influential in building up or wearing down a portion of a shelf. This is evidenced by the lack of wide shelves along the eastern coast of Florida, where the northward moving Gulf Stream strikes; the east coast of Brazil, which undergoes constant attrition by the Brazil current; and the coast of Southwest Africa, which is washed by the Benguela current. If, however, the waters are slow-moving and relatively warm, the shells of dying sea life may build up thick deposits, such as those seen along Australia's north coast.

THE CONTINENTAL SLOPE At the outer edge of a continental shelf is the shelf edge, and this marks the beginning of a steep slope, called the **continental slope** that drops off sharply down to the abyssal plain beyond. Many slopes are marked with deep submarine canyons, small mountains and hills, and a generally rough surface. The foot of a slope, at its outer margin, may be as much as 4500 to 12,000 feet (1350 to 3600 meters) below the ocean's surface. Geologists once believed that the irregular terrain of the slope was due to chemical decomposition of the materials of the slope by the ocean waters or by seismic wave action (*tsunamis*). Now, however, they believe that it results from (1) the scouring action of *turbidity currents* (fast-moving currents that are heavily laden with sediments), (2) erosional activities that occurred when the ocean was lowered during the ice ages, thus exposing the shelves and slopes, and (3) faulting along the continental coastlines. It is probable that all three processes should be considered. In addition, that some of the slopes have very steep faces indicates that landslides are common along many of the shelf edges. Good examples of this can be seen in the slopes off Cuba and Ceylon, which have slope angles of 45 degrees.

Wave Action

In chapter 6, we studied the movements of ocean waters and learned why and how ocean currents and wind waves are generated. Now we shall consider the effect of currents and waves on the landmasses of the world. Ocean currents are important to mankind as an influence on the climate and weather and are also important as sculpturing agents.

How do currents compare to wind-produced waves as sculpturing agents? The movement of ocean waters along and against a coast may be localized—that is, limited to specific areas—but if the current is fast enough, the depth of water shallow enough, and the wind-born waves powerful enough, the outer margin of the continents (or any landmass) will be greatly affected. If these conditions prevailed, erosional activity would be pronounced. If, on the other hand, the current moved slowly, the waters were deep, and the waves were gentle, depositional activity would result.

WAVE REFRACTION AND THE LONGSHORE CURRENT   There are water movements that occur along a coastline independent of the major currents that we considered earlier. These are currents born of waves. That is, waves rolling in toward a shoreline will follow the direction given to them by the wind that created them. As the waves advance toward the shoreline, the water behind them piles up in the shallower depths and the waves crest and break upon the beach in lines parallel to the beach. Waves usually do not approach a shoreline that way, however. If the waters are deep, a wave advancing at an angle of about 45 degrees would probably strike the beach at about that angle. If, however, the water depth were shallow across the shelf, the waves would tend to "bend around" as they advanced. Figure 14–3 illustrates what happens to such a wave advance. The circular wave motion, generated in deep water, gradually weakens as the wave travels a great distance. When it moves across the comparatively shallow waters of the shelf, the wave motion is greatly reduced, and the wave begins to slow along its beach side, while the seaward extension continues at the original speed. The axis of the wave then swings around until the crest is almost parallel to the beachfront. This process is called **wave refraction.**

As the waters surge up (*swash*) the sloping beach, they are not advancing parallel to the beachfront, but when they are pulled back, they move at a right angle to the beach. This produces a zigzagging motion that moves both the water and sediments picked up along the beachfront. This down-the-shoreline movement is termed the **longshore current** and is responsible for the transportation of considerable quantities of sand and other sediments along the coast. If a submarine canyon were in the way, the sand would pour into the canyon and be lost to the beaches further down the coast.*

* Encyclopedia Britannica Films has produced an excellent 16 mm movie, entitled "The Beach: A River of Sand," on the effect of the longshore current on beaches. It tells the story of the rise and decline of beaches owing to the robbery by submarine canyons and the cutting off of new sand supplies as dams are built across rivers that once carried sediments to the beaches.

FIG. 14-3 Wave Refraction and the Longshore Current.

<table>
<tr><td>⌄⌄⌄⌄></td><td>Wave refraction</td><td>- - - - ⟶</td><td>Direction of longshore current</td><td>∖ ∖ ∖ ∖ ∖</td><td>Advancing waves</td></tr>
</table>

A surf zone phenomenon closely related to the longshore current is the **riptide** (also known as *rip current,* or *feeder current*). Riptides are fairly common along beaches that have a relatively steep foreshore (beachfront). When the wave rushes back into the surf zone, it works its way out through a groove (or throat) gouged in the sandy floor to a position beyond the breakers. There, it swirls about in a circling motion at speeds of as much as 10 miles (16 kilometers) per hour. Strong swimmers often find it difficult to free themselves from the rapidly whirling waters.

TIDAL CURRENTS   It may be difficult to think of the rise and fall of water from high to low tide as a force strong enough to cause erosion or deposition, yet this does happen. An incoming tide, the change of low water to high water, can carry marine sediments onto a beach in appreciable amounts. The outgoing tide can reverse the process as it carries away previously deposited sediments. Such erosion and deposition is a minor force, and it takes considerable time for it to alter the shape of a continental shelf.

More pronounced changes occur in a river's estuary, such as Canada's Bay of Fundy (located on the continental shelf between Nova Scotia and New Brunswick). Here the tidal onrush of water raises the level some 50 feet (15 meters) at high tide, and the speed of the incoming tidal wave may approach 14 miles (22 kilometers) per hour. At low tide, the boats of the fishermen lie in the mud, tied to a buoy that also lies in the mudflats. Then comes the new high tide in the form of a steep-fronted wall of water, and the boats are summarily floated, swung about to tug at their lines. The fishermen hurriedly take to the open sea, knowing that they had better be back at anchor before the next low tide.

Why are such tidal changes so extreme in estuaries? The reason is that the rising waters are compressed between narrow banks, forcing the water to move at a much faster speed. Somewhat similar is a phenomenon called a **tidal bore,** which results when a rising tide is blocked in an estuary by mounds of sands (sandbars). The water builds up pressure until it finally breaks through and races upstream.[2]

Some areas are noted for having exceptionally high tides, with serious damage resulting, such as on the continental shelf between England and the Netherlands. In February 1953, a very high tide invaded the Low Countries, submerging more than 6000 square miles (15,600 square kilometers) of land and drowning about 1800 people.[3] There are other areas where the tidal range is so low that it is scarcely noticeable, such as in the Mediterranean Sea, the Baltic Sea, and the Gulf of Mexico.

THE SURF ZONE   While longshore currents, tidal bores, and tidal currents do perform some erosional work, it would appear that the major erosive force is produced by wind-created waves that endlessly attack the beaches of the world. This is accomplished by one or both of two erosion processes: hydraulic action and corrasion.

Hydraulic action refers to the actual pounding of a wave upon an object in its way. An advancing wave can exert tremendous pressure as water piles up behind a cresting wave and then drives it forward like a sledgehammer. Robert C. Cowen tells of storm-tossed waves that struck the Tillamook Lighthouse on the east coast

(a)

(b)

The Bay of Fundy, Nova Scotia. There is a dramatic change from low tide (a) to high tide (b) as the ocean's waters rush through Minas Channel, pass Cape Split, and move up the river at Parrsboro.

of the United States and sent a 135-pound (60-kilogram) rock so high into the air that it came down through the roof of the 91-foot (27-meter) tall structure.[4]

Corrasion (another term for *abrasion*) refers to the sandpapering effect created when the advancing wave picks up sand or gravel and abrades the sea cliffs and other surface features of the shoreline and the shelf.

Landforms of Wave Erosion

Figure 14–4 illustrates some of the more common landforms resulting from wave erosion. All such forms were sculptured out of the original landmass or out of rock layers laid down at a later period of geologic time.

Sea terraces are the remnants of ancient shorelines that have been uplifted to their present height above sea level, and many of these exhibit landforms of wave erosion that occurred millions of years ago. The fronts, or faces, of these terraces

FIG. 14-4   Landforms of Wave Erosion.

Erosive Power of Ocean Waves. Shown in this photo of a portion of the coastline of France are a wave-cut bench, sea cliffs, an arch, and an isolated stack.

are called **sea cliffs.** They offer proof of the erosive action of the waves. Sometimes heavy surf action grinds out a horizontal groove, termed a **notch,** at the base of a sea cliff. If the notch is enlarged sufficiently that the cliff front is no longer supported, it may fall, thus pushing back the face of the cliff. This is called the "retreat of sea cliffs."

Sometimes you will find a variety of landforms scattered along the front of an old, eroded sea cliff: **arches** (protruding masses of resistant rock that have had their undersides hollowed out by wave action), **stacks** (remnants of resistant rock after the bridge portion of an arch has collapsed), and **sea caves** (which result from wave action against easily eroded or fractured rock beds). The continued pounding of waves into the confined, hollowed-out space can break the rocks apart, thereby enlarging the sea cave. Eventually, the roofs of some of these caves are weakened by the wave action, and the overlying rock collapses into the cave, forming a canyonlike cleft. Some of the Hawaiian Islands have an interesting variation of a sea cave: a *lava tube* (the result of lava cooling around a fast-flowing stream, with the stream thereafter exiting from the hardened lava). When waves surge forward and enter the open end of the tube, they eventually break through somewhere along the line and spout into the air like a geyser. This gives them their name, *spouting horns*.

The beach fronting the sea cliff is often termed a **wave-cut bench,** or **abrasion platform,** and here, the layers of solid rock have been leveled and eroded by forward and backward action of the surf, leaving behind a relatively flat, benchlike surface that is usually covered with rock debris. It may be submerged during high-tide periods and fully exposed at low tide. Material taken from the abrasion platform is mixed with sands brought in from the shelf beyond to create a **wave-built terrace.**

378

Landforms of Wave Deposition

Figure 14–5 shows a typical beach scene, with landforms resulting from the deposition of sediments by wave action. Most of these forms are, of course, composed of sand.

Beaches can be produced by erosional or depositional processes, depending on any of several factors: the power of the surf, the structure of the underlying rock, the angle the shoreline presents to the incoming waves, and the level of the sea. The surface covering of any beach usually comes from somewhere else, brought down from inland areas by streams and rivers, and its composition would depend on the rocks from which the material was derived. Southern California beaches are light

FIG. 14–5 Landforms of Wave Deposition.

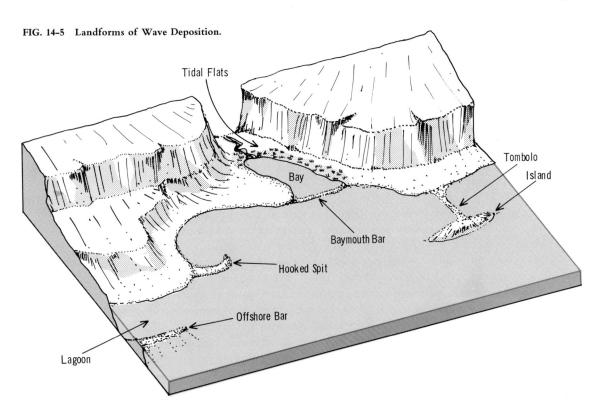

grayish, derived from weathered granite; the island of Hawaii has white coral-derived sand on the east, black sand from weathered lava on the southeast, and green sand from weathered olivene on the northwest. Beaches along England's south coast are often referred to as *"shingle" beaches,* because they are built of pebbles and cobbles that are flat, thus giving the beach the appearance of a shingled roof. The beach south of Ensenada, Baja California, is littered with large cobbles and small boulders that are laid out in parallel rows over an eroded granite base.

Barriers include a number of landforms that always seem to have the shape of a bar—that is, their length is greater than their width. **Sandbar** is the most common term for a mound of sand located along a shoreline, but physical oceanographers have assigned these formations more representative names. An **offshore bar** is created by incoming waves that remove sand grains from the beach during high wave conditions (stormy seasons), drag them in the backwash to a position away from the shore, then deposit them in a pile that is parallel to the beachfront. Because the longshore current often works to extend these bars for a number of miles, they are sometimes called *longshore bars.* Some interesting examples of offshore bars that have become islands and serve as protective barriers to the coastline can be seen along the eastern coast of the United States and on the Texas coast of the Gulf of Mexico. Fire Island stretches for 27 miles (43 kilometers) off the coast of Long Island and has been described as a place that continually experiences "the ceaseless sound of booming surf, whistling wind-driven sand, and the crying of gulls."[5] It resulted from the interaction of the tides, ocean currents, and wind waves on the materials of the continental shelf. The island is perpetually altering its face and is, in fact, being driven in a westerly direction as its windward side is eroded and its leeward side is being added to by the elements.*

Padre Island in Texas, also a barrier composed of sand, is about 80 miles (128 kilometers) in length. It is just one of several barrier islands that follow the coastline of Texas from Galveston around to Mexico. The barrier island that is the home of the city of Galveston is more properly termed a **baymouth bar,** because it was created by longshore currents at the mouth of Galveston Bay. This bar comes very close to joining the southwesterly jutting Bolivar Peninsula. The body of water separating Galveston and its island from the mainland is a typical **lagoon.** In order to provide access by means other than boats, large causeways (elevated highways) were constructed across the waters.

Spits are long, narrow sandbars, created by the depositional activities of a longshore current flowing past a cape or promentory that protrudes from the mainland. As the current curves around the end of the landmass, its velocity is slowed and deposition begins. One end is affixed to the land, the other points downcurrent. From above, the spit does look very much like the cooking tool used to hold meat over a fire. If the longshore current sweeps around a promentory or headland into a bay, the deposition activities will follow, forming a curving sandbar that is called a **hooked spit.**

---

*To learn more about the world's beaches, barrier islands, and other features bordering the oceans, especially such interesting places as Fire Island, New York, and Padre Island, Texas, see: Robert and Seon Manley (1968), *Beaches: Their Lives, Legends, and Lore* (Philadelphia, Penn.: Chilton).

(*a*)

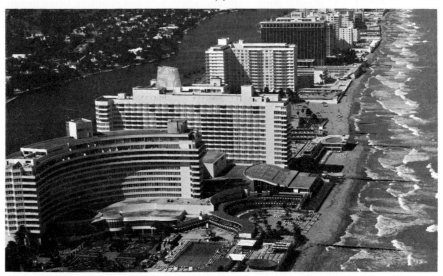

(*b*)

Barrier Islands and Sandbars. (*a*) Offshore bars and long spits act as barriers to the beaches along North Carolina's coastline. This photo (taken at the end of summer) shows the numerous sandbars created by relatively gentle summer waves within the lagoon. (*b*) Hotel Row in Miami Beach, Florida, occupies a large offshore bar that might well be called a barrier island.

Spits. The shoreline of Martha's Vineyard, Massachusetts, displays a variety of spits, including a hooked spit that has almost completely encircled its lagoon.

Tombolo is the name given to a ridge of sand that connects an offshore island to the mainland. When advancing wind waves and the longshore current are confronted by such a landmass, the sediment-laden waters must curve around the island, thus having their velocity slowed until the sediments can no longer be supported. Deposition begins along the beachfront directly behind the island and in time will reach out and become fastened to the island. The most famous tombolo is the one connecting the Rock of Gibraltar with the mainland of Spain.

Shorelines

Shorelines differ greatly from place to place around the landmasses of the earth. There have been many attempts to classify them, usually resulting in more confusion than enlightenment.

Some shorelines have been exposed to the various erosional activities for vast periods of time. Mountains along a coast may have been worn down to low hills while streams and rivers were scouring out wide valleys and causing the formation of large alluvial fans. If this coastline were submerged by a sinking landmass or the rising elevation of the sea, the new coastline would be deeply indented with many estuaries and bays. It would be an irregular shoreline.

On the other hand, on a shoreline where deposition had been occurring all along the continental shelf, thus creating a fairly smooth and gently sloping plain, if the shoreline were upraised or the level of the sea lowered, the result might be quite different. There would be hills and valleys, along with wave-cut benches and sea terraces, but they would be in the background. The forepart of the shoreline would look much like the piedmont alluvial plains seen along the base of great mountain ranges.

Complications occur, and it is sometimes difficult to distinguish a **shoreline of emergence** from one of submergence. For example, the shoreline along Monterey, California, is extremely rugged, deeply indented by clefts and bays, and the beaches are not all long and wide. It is a shoreline that has emerged, but it has all the characteristics of a **shoreline of submergence.** Why? Because this shoreline is situated on the eastern edge of the shifting Pacific plate, and the tremendous diastrophic forces operating there have crumpled the land as the coastal ranges were uplifted. Landslides crashed down from the newly emerging mountains and poured into the oceans, creating powerful turbidity currents that swept sand and other sediments before them as they gouged out submarine canyons and roughened the surface of the continental shelf.

Glaciated shorelines are not so confusing, for they are the result of processes that are still occurring. Along the coasts of Alaska and Norway, the shoreline is extremely rugged, with an almost endless number of bays and deep valleys. A moving glacier does not have to conform to a baselevel; it can gouge as deeply as its weight and pressure will allow. A glacier moves down from the higher elevations, carving a valley as it goes. When the glacier confronts the ocean, it moves on out into the water, and throughout its span of existence, the front edge, or terminus, breaks off in pieces (bergs). When the glacier begins to melt faster than it can be resupplied from its source, it begins to recede back up through its valley, and the ocean follows. The glacial trough is said to be "drowned," and the result is an intrusion of the ocean into what looks like a common bay. This is called a *fjord* (fiord).

Coral shorelines are the result of the activities of living organisms, such as coral, algae, and shellfish. These organisms thrive in relatively warm waters of 68°F (28°C) or higher and make their home on the upper level of an offshore bar, volcanic island, or sea mount. Coral growth invites other organisms to live in the growing reef, and as organisms die, leaving their shells to build up the formation, other organisms take over and the cycle continues.

Charles Darwin evolved a theory to explain why we find coral **fringing reefs** (coral growth and debris surrounding a submerged volcanic island) scattered throughout the South Pacific Ocean. Knowing that coral cannot survive in cold,

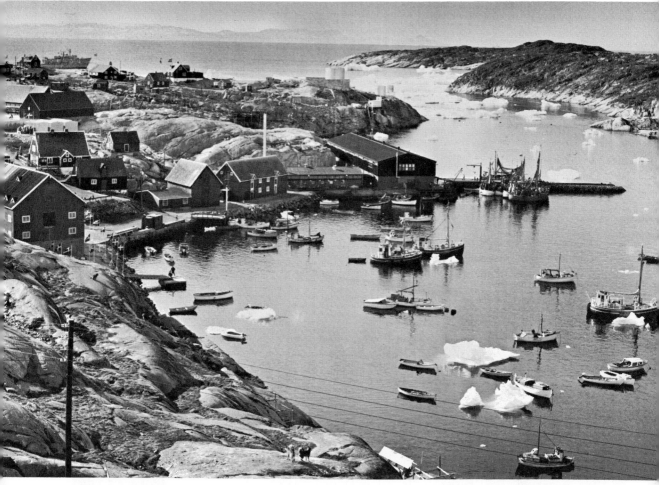

Glaciated Coastline of Denmark. Note the bedrock, which has been laid bare by advancing glaciers, and the fjord, now used as a harbor.

deep waters, he surmised that the reefs must have had their beginning when the island was at or near sea level. When the islands subsided, the coral continued to grow, and the reef maintained its position relative to the level of the sea.

There was considerable controversy over Darwin's theory, with some geologists and oceanographers insisting that the coral grew when the sea level was lowered during the great ice ages. Time and research has proved most of Darwin's coral-reef theories to be correct.*

*To learn more of the "coral-reef controversy," see: Robert C. Cowan (1969), *Frontiers of the Sea,* rev. ed. (Garden City, N.Y.: Doubleday), pp, 112–119; and Francis P. Shepard (1973), *The Earth Beneath the Sea* (New York: Atheneum), pp. 185–207.

Swain's Island off the Eastern Coast of Australia: Coral Atoll and Fringing Reef.

Barrier reefs are unlike fringing reefs in that they stretch for great distances along a coastline. The Great Barrier Reef off Australia's east coast extends for some 1200 miles (1920 kilometers) and is the most famous of all coral barrier reefs. Fringing reefs are called **atolls,** if the mountain on which they are built is below sea level. The South Pacific is dotted with these atolls, which have a special kind of beauty: deep, clear blue waters in a calm lagoon, surrounded by weathered coral deposits where coconuts have taken root, providing an existence for humans who prefer life on an idyllic tropical isle.

Volcanic shorelines occur wherever lava flows have reached the surf zone. Here, one finds great black cliffs and terraces that tower over the breaking white surf. In some places, the beaches are of black sand, derived from the action of the surf on the lava, and in other areas, the shoreline has the appearance of a gigantic mass of coiled black rope.

Black Sand Beach, Kalapana, Hawaii.

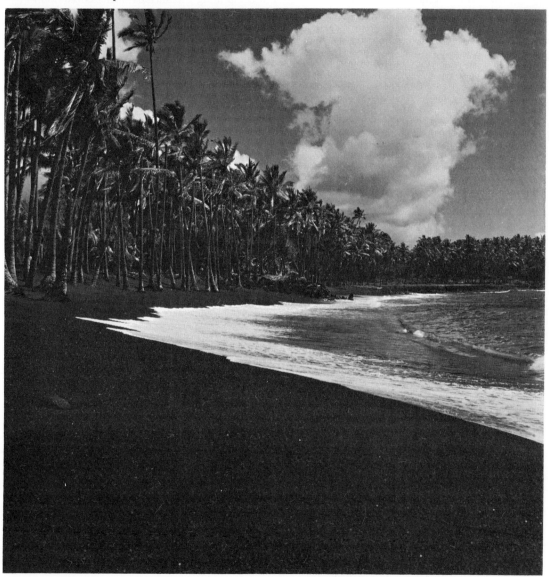

Review and Discussion

1. Define the term: *continental shelf.* Where are the widest shelves to be found? How did they originate?

2. What is the relationship between the shelf edge and the continental slope?

3. Why are continental shelves important to people today?

4. What is the effect of turbidity currents in sculpturing the surface of the continental shelves and slopes?

5. What is the relationship between wave refraction and the creation of longshore currents.

6. What is a riptide?

7. Explain why tidal currents can be an effective erosional force.

8. Explain the difference between the terms *hydraulic action* and *corrasion* with reference to wave erosion.

9. Discuss the landforms of wave erosion, specifically: sea cliffs, arches, stacks, and wave-cut benches.

10. What is a spouting horn?

11. How are beaches formed?

12. Describe the different barriers created by wave deposition. What is the difference between a baymouth bar and a spit? What is a tombolo?

13. Why is it difficult to classify shorelines of emergence and submergence?

14. Describe the formation of a glaciated shoreline.

15. What is the difference between a fringing reef and a barrier reef?

16. Describe a volcanic shoreline.

Notes

[1] Robert C. Cowan (1969), *Frontiers of the Sea,* rev. ed. (Garden City, N. Y.: Doubleday), pp. 19–20).

[2] Leonard Engel (1961), *The Sea* (New York: Time-Life Books), p. 94.

[3] Leon Bertin (1972), *The New Larousse Encyclopedia of the Earth* (New York: Crown), p. 110.

[4] Cowan, p. 134.

[5] Robert and Seon Manley (1968), *Beaches: Their Lives, Legends, and Lore* (Philadelphia, Penna.: Chilton), p. 324.

Additional Reading

Flanagan, Dennis, ed. (1969). *The Ocean.* San Francisco: W. H. Freeman.

Bascom, W. (1964). *Waves and Beaches.* New York: Doubleday.

Carson, Rachel (1961). *The Sea Around Us.* New York: Oxford University Press.

Cromie, W. J. (1962). *Exploring the Secrets of the Sea.* Englewood Cliffs, N.J.: Prentice-Hall.

Darwin, Charles (1959). *The Voyage of the Beagle.* Abridged and ed. Millicent E. Selsam. New York: Harper & Brothers.

King, C. (1972). *Beaches and Coasts.* 2d ed. New York: St. Martin's Press.

Maury, Matthew Fontaine (1963). *The Physical Geography of the Sea and Its Meteorology.* Ed. John Leighly. Cambridge, Mass.: Harvard University Press.

15 Engraved by Glaciers

Glaciers mean different things to different people. To some, glaciers represent water: about 2 percent of the earth's total water supply and close to 10 percent of the fresh water. To others, glaciers represent an ever-present threat—for example, the citizens of isolated villages in Norway or Alaska, where a gleaming white wall of ice poises like the blade of a bulldozer before their homes. Yet, to still others, the results of glacial activity—such as in the magnificent valleys of Yosemite and Glacier national parks in the United States and Banff and Lake Louise in Canada—are sufficient evidence that some portions of the earth must endure the tortures of an icy hell before they can reign in all their majestic glory.

Curious individuals have long pondered the sometimes delicate, sometimes bold and sharply etched peaks, ridges, and valleys of the Alps, the Hindu Kush, or the Caucasus. Whose hands fashioned those engravings? And what of the huge boulders that litter the landscape from the foothills of the Alps northward to the Baltic Sea? Had playful or angry giants tossed them there? Had a great flood transported them from the north? Were they vomited out of the earth's interior? And what of the towering granite cliffs that shone like polished metal under the noonday sun? Who or what made the deep scratches on high valley walls, scratches that were parallel to the valley floor?

The centuries rolled by, and although many were satisfied with the answers that evolved, others could not contain their sense of frustration. Then, in 1836, a young Swiss zoologist arrived on the scene: Louis Agassiz.

Agassiz might well be called the "father of modern glaciology," because through his scientific reasoning, the answers to all those questions finally began to take form. He had studied the glacial activity of the high valleys of the Alps, and when he journeyed through the rolling landscape of the North European lowland, he noted that the huge boulders, along with the other earth materials and scattered rock debris, were remarkably similar to those found in the high alpine valleys. He evolved a theory that there once had been a great glacier covering the land, a glacier that would make the glaciers of the Alps seem tiny by comparison. He theorized that the northern regions of Europe and Asia had suffered through a number of ice ages for a greater part of earth's history, and each time the ice sheets moved out, they scoured the land and transported rocks of all sizes, from sand grains to gigantic boulders.*

While Agassiz's views generated considerable discussion and opposition, his theories have prevailed. Glaciology (from the Latin *glacies,* "ice") moved from a purely hypothetical level to one of scientific analysis, based on evidence.

Aside from gaining knowledge, has mankind benefited materially from a serious study of glaciers? The answer is an unqualified yes. For the proof, let us proceed with an investigation into glaciers, past and present.

---

*To learn more of Louis Agassiz, see: Carrol L. Fenton and Mildred A. Fenton (1952), "Agassiz of the Ice Age," *Giants of Geology* (Garden City, N.Y.: Doubleday), chap. 10, pp. 111–123; Charles Ogburn, Jr. (1968), *The Forging of Our Continent* (New York: American Heritage), pp. 116–117; Ruth Moore (1971), "Agassiz: Ice!" *The Earth We Live On* (New York: Knopf), chap. 7, pp. 132–156.

The Formation of Glaciers

When moisture is present, along with extremely low air temperatures, a number of different ice forms are created, such as frost, hail, sleet, permafrost, snow, and glaciers.

Frost can occur almost anywhere in the world, whenever moisture-laden air close to the ground is chilled below the dew point. Hail and sleet are frozen water droplets that may have started to precipitate as raindrops, but they, like frost, are short-lived.

The permafrost regions bordering the Arctic Ocean are lands of underground ice, where the ground 6 to 9 inches (15–23 centimeters) below the surface is permanently frozen. Although these regions offer a unique and sometimes exciting landscape, they are not glacier-producing areas today. They may receive up to 25 inches (63 centimeters) of precipitation annually, but most of that arrives during the short summer, when the top layer of the ground thaws. Glaciers are born from the accumulation of snow, but not all snowfields produce glaciers, because they may lack one or more of the formation controls.

FORMATION CONTROLS   The formation and growth of glaciers is determined by the operation of the hydrologic cycle. Glaciers can form only where the accumulation of snow remains throughout the year, which happens when the annual snowfall exceeds any losses resulting from evaporation or melting. Although the amount and type of precipitation are important, so are the temperatures of the air masses and the land over which they move. Some mountains in the Rockies may receive as much as 40 feet (12 meters) of snowfall during the winter months, but when summer finally comes to an end, only small patches of snow remain. The continent of Antarctica receives only about 6 inches (15 centimeters) of snowfall annually, yet scientists who participated in the International Geophysical Year (IGY) discovered that the ice sheet covering the land and extending out into the seas was actually growing.[1] Why do these regions differ so greatly? The answer lies in the differences in temperatures and the amount of precipitation.

Temperature, a very important snowfall control, is determined by altitude and latitude. Snowflakes or snow crystals will precipitate wherever the air temperature is cold enough for the water molecules to freeze as they condense. The snowflakes may melt if they fall through warmer air, or they may remain in a solid state until they strike the ground. There, if the temperatures are warm, they might revert to liquid form. Whether the fallen snow remains to become a part of the snowpack depends on the altitude of the snow line.

The term **snow line,** which varies with the latitude, means the lowest elevation at which a permanent snowfield or pack can perpetuate. The snow line is sometimes called the *tree line,* because it is the upper limit of tree growth. On Tanzania's Mt. Kilimanjaro the elevation of the snow line is about 15,000 feet (4500 meters); in the Alps, 8500 feet (2550 meters); and along the northern borders of Alaska, close to sea level. This is not to imply that the snow line never varies in altitude from place to place along a specific parallel of latitude; it is also influenced by the moving and moist air masses in the region, whether the prevailing winds are cool or warm, the receipt of insolation, and the variations in daily, monthly, or annual temperatures. A storm moving out of the polar front may drop great quantities of snow on the Rockies, the upper Great Plains, and the lowlands of Iowa and Illinois. How long does the snow cover last in each of these areas? Obviously, a single storm, even a severe one, will not alter the level of the snow line permanently.

Snowfall is dependent on the air temperature, but it is also determined by the amount of moisture present when precipitation occurs. If a cold, dry air mass rises over a mountain range or a warm, dry air mass sweeps up from the subtropics toward the high latitudes, there will be no precipitation. However, if a moist air mass follows either of those paths, precipitation may result, and if the air temperatures are at or below freezing, sleet, hail, or snow will result.

GLACIAL STRUCTURE   Assuming proper freezing conditions at ground level and a plentiful supply of falling snow, the snowflakes or crystals will begin to accumulate in an ever-thickening mass. The greatest depths will occur in valleys, rather than on mountain ridges, for the simple reason that a heavy pack of snow would slide off a ridge. So, in high valleys, the snowpack deepens from the head (the highest elevation) down to the snow line (Fig. 15–1). This is called the **zone of accumulation.** As the snow accumulates, the pressure on the bottommost layers increases. The crystals begin to break up, the snowflakes begin to melt under the pressure and then they freeze. As this is repeated again and again, layers of ice crystals form, and this impacted mass is called a **firn.**

When sufficient snow accumulates, gravity takes over, and the newly formed glacier begins to move slowly down the slope. When it descends below the snow line, the front edge, or foot, of the glacier begins to deteriorate as it loses moisture. If the underside of the icy mass moves across warmer ground, it begins to melt and thus releases streams of **meltwater.** If the upper edges encounter temperatures above freezing, surface thawing begins and moisture is lost through evaporation and melting. Sometimes the snow passes directly from a solid to a gaseous state, without going through the process of liquifying. This is termed **sublimation.** This portion of the glacier is called the **zone of ablation** (from the Latin *ablatus,* "to carry away"), or the **zone of loss.** Further deterioration can occur if a glacier enters a body of water. Losses will result from wave action and the warmer temperatures of the water, and some sections of the glacier (for which the water does not provide enough support) will break off, to float away as icebergs or ice floes.

GLACIAL MOVEMENTS   As the firn continues to increase as it accumulates more snow, the newly forming glacier may begin to move. Whether it does or not depends on a number of factors, such as the weight of the impacted snow, the steepness of the slope, and the terrain over which it must pass. Observers of glacial landscapes tell us that a mountain glacier must build to a depth of almost 66 feet (20 meters) before there can be any notable movement, and polar glaciers might remain stationary even when they have a depth of over 200 feet (60 meters).[2] Glacier movement begins with the lower portion stretching out, but the force is felt throughout the entire mass. Ice at the head of the glacier gradually pulls loose from rock wall, and

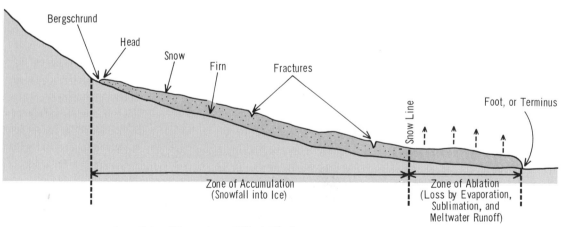

FIG. 15-1 Formation of an Alpine (Mountain or Valley) Glacier.

a crevasse (called a **bergschrund**) opens. The glacier oozes down the slope much like a river, although considerably slower and with less internal turbulence.

The speed of movement will vary, of course, depending on the factors listed above as well as on the time of the year, temperatures of the air and the ground, and whether the pathway is straight or has many curves. Some glaciers speed up during the winter, because of the increased weight added by new snowfalls; some speed up during the summer, because friction is reduced by heavy losses in the zone of ablation. Some glaciers advance rather rapidly, moving at speeds of about 100 feet (30 meters) per day, which amounts to about 7 miles (11 kilometers) per year. Others appear to be entirely motionless. It is difficult to detect the motions of any glacier, but time-lapse photography enables us to "see" the flowing movements.* In this method, with the camera aimed at a section of the glacier, a foot of film, for example, might be exposed at noon each day. The filming might cover a period of three months, and when the film is processed and shown, the mass of ice, which appeared to be stationary, becomes a flowing "river" of ice. You see it flow over a ridge of bedrock, fracturing into crevasses as it goes, sending huge blocks of ice tumbling onto the glacier's surface.

*Encyclopedia Britannica has produced a remarkable film that uses the time-lapse photography technique to show the movement of a glacier down a valley. It is entitled "Evidence for the Ice Age."

Zone of Accumulation. In this view of Boston Glacier in Cascades National Park, northeast of Seattle, Washington, note the heavy compaction of snow (firn) in the zone of accumulation.

Zone of Ablation. A stream of meltwater issues from the terminus of Franz Josef Glacier, Westland, New Zealand.

As for movements within the glacier, the rock walls and floor of the valley tend to offer great resistance; thus the upper layers and the center of the glacier move more rapidly than the bottom or sides.

Classification of Glaciers

Because glaciers come in many different sizes and shapes and can be found at varying altitudes and latitudes, an uncomplicated classification system was devised that set up two major categories: Glaciers that form in mountainous regions in high valleys are termed **alpine glaciers** (also called **mountain** or **valley glaciers**), and glaciers that cover vast expanses of land are termed **continental glaciers**

ALPINE GLACIERS   Alpine glaciers are the smallest glacial forms and originate in mountainous regions at, generally, high elevations. Their sculpturing efforts produce landscapes that are visually exciting, as anyone can attest who has driven down the Million Dollar Highway from Grand Junction to Durango, Colorado, stood on a hilltop in Alma-Ata, Central Asia, and gazed in awe up at the glaciers dotting the steep face of the Tien Shan, or hiked through the valleys near Banff and Lake Louise in western Canada.

How does such fantastic scenery come into being? When you think of the weight of the snow and ice pack confined in the narrow channel of a V-shaped, youthful stream valley, it becomes apparent that the pressures generated must be exceedingly powerful, especially against the rock walls of the valley sides. This pressure forces rocks that may already be jointed and fractured to break up into even smaller pieces. Moisture from the glacier seeps into the cracks and crevices and, if freezing occurs,

The Tien Shan Mountains.   Glaciers and snowpack are the perpetual cover for these mountains that serve as part of the barrier between the Soviet Union and China. The "uniform of the day" for the frontier guards includes both warm clothing and skis.

the rocks will be plucked from the valley walls and floor as the glacier moves on. This process, termed **plucking,** continues throughout the entire length of the glacier as long as temperatures are low enough to permit freezing. These jagged rocks, frozen to the glacier and sticking out like the spikes on golf shoes or the grains of sand on sandpaper, perform well as erosional agents as they scour and scrape the valley. This act of "sandpapering" is called **abrasion.** By these two processes great quantities of rock materials are forcibly removed from the glacial valley, eventually to be deposited in some distant place.

Erosional landforms of alpine glaciation Although a glacier may come into existence, then fade away because of the change in the region's climate, we can see the evidence of its former presence in the landforms left behind. Alpine glaciers perform both erosional and depositional activities, but Fig. 15–2 shows that these high mountain glaciers are more to be noted for their erosional activities.

Erosional landforms have their beginnings when the firn develops within a valley at normally high elevations. The valley usually is a youthful stream valley that is narrow and V-shaped. When the glacier begins to move, plucking and abrading as it goes, the valley becomes rounded into a U-shaped profile. The head, or top, of the valley assumes an almost circular shape (much like the end of a bathtub), which earned it the name of **cirque** (from the French word meaning "circle"). When a glacier melts away, the only objects in the cirque will be a few scattered boulders deposited by the glacier and a shallow pool or lake (called a **tarn**).

As the glacial valley widens into a rounded **trough,** the erosional activities encroach on the valley sides, cutting away at the mountain spurs, or ridges, on either side. If two glaciers are operating in adjacent valleys, the ridge between will become sharply etched, almost knifelike. These are termed **arêtes** (from the French word meaning "fishbone" or "sharp ridge"). If two glaciers are situated with their heads back-to-back, the plucking created by the constant activity at the bergschrund, or head of the cirque, the pulling away of the top of the glacier alternating with refilling and freezing of the crevasse, may cause the two cirques to become joined at the top of the ridge. If this happens, a section of the arête will be cut away, leaving a **col** (from the Latin *collum,* "neck"), or mountain pass, through the ridge. When several glaciers are all at work on a mountain peak, carving their cirques upward, creating arêtes on all sides, the resulting sharply chiseled peak is called a **horn,** because it bears a striking resemblance to an animal's horn. In Switzerland, the tallest horn in an area is called the *matterhorn* ("mother horn").

The rounded, U-shaped valley created by the passage of a glacier is called a **glacial trough.** If such a trough is formed in a valley leading to the shoreline and its base is below the level of the sea, the waters of the ocean will enter the trough as the face of the glacier melts and recedes. The resulting "drowned" glacial trough is called a **fjord** (*fiord*). The areas most noted for the number and picturesqueness of their fjords are the heavily indented coastlines of Alaska and Norway. Some of the Norwegian fjords reach inland as far as 120 miles (192 kilometers).[3] Fjords predominate in these high-latitude regions (as well as along the coasts of Greenland, Scotland,

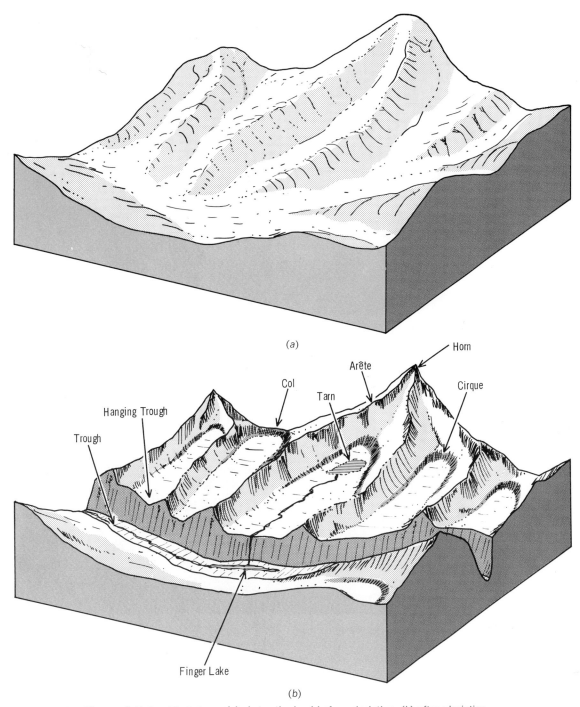

FIG. 15-2 **Landforms of Alpine Glaciation.** (*a*) shows the land before glaciation, (*b*) after glaciation.

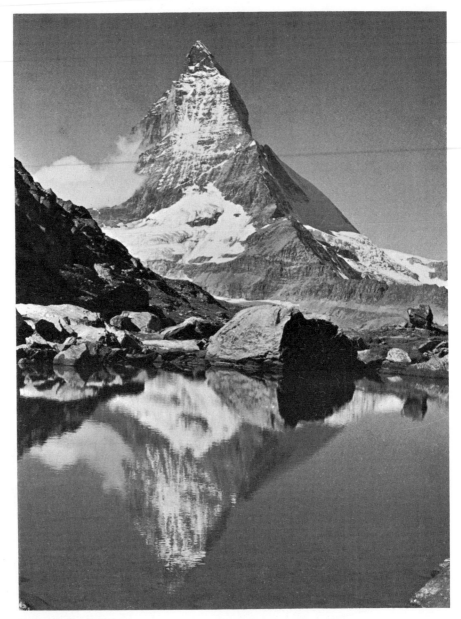

Switzerland's Matterhorn.

and Chile, among others) because of the climate and the mountainous topography sculptured during the early ice ages.

If another glacier cuts across the foot of a glacial valley and erodes the land away, the tributary glacier will be left hanging high above the intruder's newly forming valley. The result is termed a **hanging trough**. A moving glacier often will scour out a long depression on the floor of its valley, and when the glacier has receded, the narrow gouged-out cavity may fill with water issuing from the receding glacier, and a **finger lake** is born.

What happens when a glacier encounters a ridge of resistant metamorphic or igneous rock? Does the glacier flow around the obstacle? Although a glacier flows somewhat like a river, there, the resemblance ends. You may think of ice as being brittle and easily shattered, as not having the strength or resistance of a rock, but consider the power of a great mass of ice meeting an exposed volcanic dike. The glacier flows over it, abrading the *stoss* (from the German, "thrusting") side as it climbs, and plucking pieces off the lee side as it moves on (Fig. 15–3). Looking down on a glacial trough dotted with these rounded rock forms, you might think they resemble a herd of grazing sheep. This is why the Swiss named these erosional features **roches moutonnées**, which means "sheep rocks."

Glaciers sometimes move through areas of highly resistant bedrock, with the rocks plucked from places upstream scraping at the rock walls, making deep scratches or grooves, called **striations**. If a mass of granite is polished smooth by the abrasive action of the ice, it will shine like a mirror, and this is termed **glacial polish**. These items may seem unimportant, but they do help us to determine which processes created a particular landscape or landform.

Sognefjord, Norway. The longest fjord in the world reaches 110 miles (175 kilometers) into the heart of Norway.

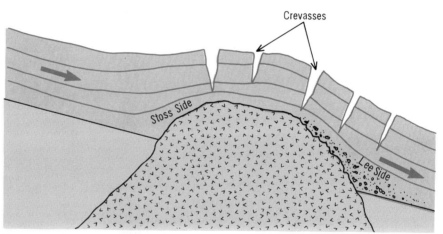

FIG. 15-3 Glacial Sculpturing. A glacier moving across an extrusion of bedrock grinds it into a roche moutonnée ("sheep rock").

Depositional landforms of alpine glaciation Depositional landforms are created by advancing as well as receding or retreating glaciers. When we speak of glacial advance, we are referring to the actual, physical forward movement of a glacier as it surges down its slope. A receding glacier is not actually moving back up its slope. It only appears to be. Glacial recession begins when a glacier no longer receives replenishment in its zone of accumulation to make up for the losses occurring throughout the zone of ablation. As pieces of the glacier break off or the ice melts, the **terminus** (front edge) of the glacier appears to retreat. If this process continues, the glacial trough eventually loses its glacier, and the only evidence of its existence is the erosional forms made by the moving glacier and the deposits left behind.

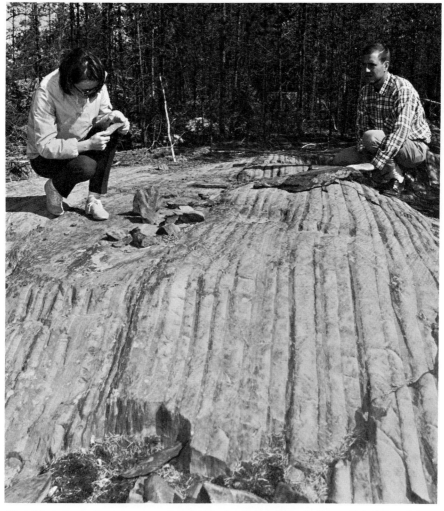

Glacial Striations. These grooves, found in bedrock near Yellowknife, capital of Canada's Northwest Territories, were carved by an advancing glacier during an early ice age.

The area over which a glacier passes is strewn with earth materials, ranging in size from silt particles, sand grains, pebbles, to large boulders (**erratics**). Such random, unsorted, and unstratified deposits are called **moraines** or **drift deposits.** The scattered debris found on the floor of a glacial trough forms a **ground moraine** or **till plain.***

As a glacier moves down its valley, rock debris and fragments fall from the valley

*The term *till plain* is generally used to refer to the ground moraine left by a receding continental glacier. The word *till* means "glacial deposits," which would apply to either alpine or continental glaciers, but the small glaciers do not produce the broad plains associated with the vast ice sheets of continental glaciation.

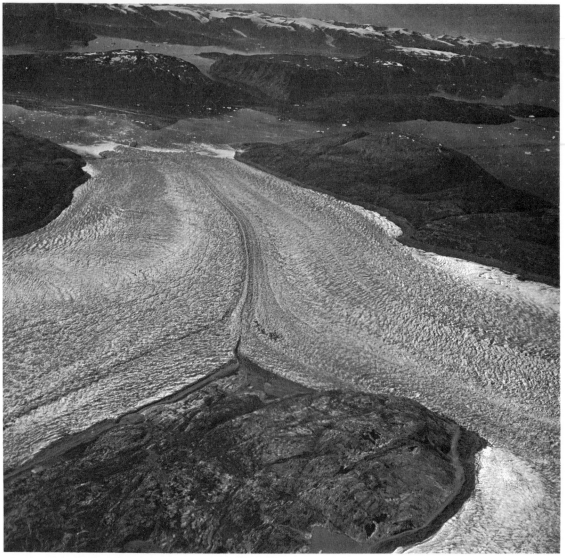

The Convergence of Glaciers. Two glaciers in Greenland join, and their lateral moraines merge into a quite visible medial moraine.

sides onto the edges of the glacier, resulting in morainal deposits called **lateral moraines.** When two glaciers extend beyond their valleys and merge in another valley, the joining of the lateral moraines produces a **medial moraine,** which shows as a dark stripe down the middle of the merged glacier.

Rock debris and soil materials carried by streams flowing over the surface of the glacier and deposited along the foot (or terminus), along with materials pushed by the glacier itself, form deposits called **terminal** or **end moraines.** If a receding glacier pauses long enough in any one location to reestablish itself (because of the increase in snow accumulation at its source), it forms a new terminal moraine before it begins melting away. One may discover a number of these **recessional moraines** in an empty glacial trough.

There are other depositional features associated with alpine glaciation, but because they are more commonly associated with the landforms of continental glaciation, we shall consider them in that section.

CONTINENTAL GLACIERS   So far, we have primarily considered the relatively small glaciers that perform their sculpturing and leveling activities in high mountain valleys. Now it is time to consider the vast ice sheets we call continental glaciers. In ages past, there were many, but today we have only two that fit within this classification: the ice cover of the Antarctic continent and the icecap over Greenland. How do these great glaciers form? The answer to that question will be found by examining the conditions that created the continental glaciers of the geologic epoch known as the Pleistocene (from the Greek *pleistos,* "most," and *cene,* a revision of *kainos,* "new," thus, "most new"), the most recent of the ice ages.

The Pleistocene Ice Age   The world's climate has changed at least four times during the past million years, probably because incoming solar radiation was blocked off by an excessively heavy layer of clouds, allowing the temperature of the atmosphere to fall until the moisture precipitated over the polar regions fell as snow. (See chapter 3, pp. 73 to 80, and chapter 6, pp. 168 to 171). As snow continued to fall, and the resulting snowpack became deep enough to form glaciers, great tongues (or lobes) of ice began to reach equatorward from the poles. Reaching out from the poles, time after time, they covered most of the northern portions of North America and Eurasia, advancing for thousands of years, then receding as the climate warmed and the ice melted, only to advance again as the climate turned colder.

Glaciologists and geologists have, however, been able to study the performance of the continental glaciers formed during the Pleistocene. Figure 15–4 shows the extent of this most recent ice age. Consider how dramatic were the effects of this period of glaciation on humans. With the level of the oceans dropping as moisture was trapped in the growing ice sheets, land bridges were exposed, enabling migrating people and animals to move from one landmass to another. Volcanic islands became mountains, standing above sloping plains that previously were part of the ocean floor. It is quite possible that the ancestors of today's Australian aborigines followed such an exposed pathway from Southeast Asia to the land "down under." It is also possible that, when the glaciers finally melted and returned their waters to the seas, the land bridge the aborigines followed became submerged, thus preventing return to their native land. The most accepted theory concerning the route taken by the ancestors of today's American Indian—the "first Americans"—is that a land bridge connected Alaska to Siberia during the Pleistocene.

All of this did not happen at one time. The Pleistocene is believed to have lasted about 1 million years, coming to an end about 10,000 years ago. During this span of time, the continental glaciers made four advances (stages) and retreats (interglacial

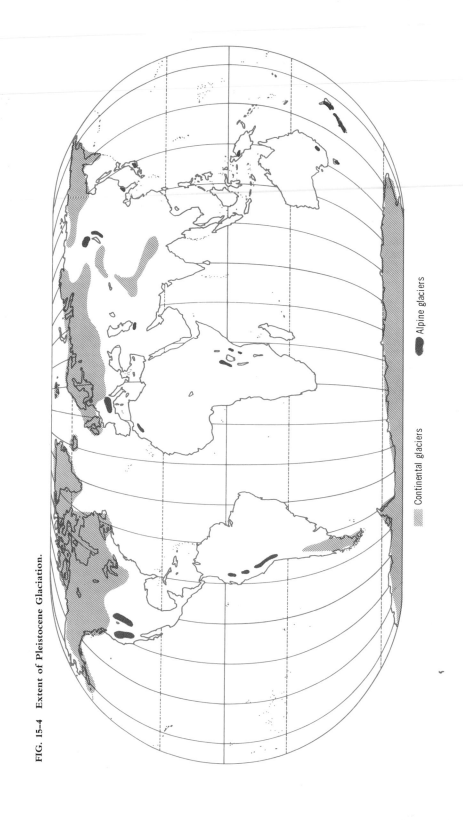

FIG. 15-4 Extent of Pleistocene Glaciation.

Continental glaciers   Alpine glaciers

covered with till deposits. When the ice melts completely, there is only a hole to show where it once was. These are called **kettles.**

Moraines are as common to continental glaciers as they are to alpine glaciers, only they are more prevalent and cover much greater distances. The same terms apply to the different types: lateral, terminal, ground, and recessional. The only exception is the medial moraine. Because a continental glacier does not operate as a single, small ice mass that can unite with another, but moves as broad fingerlike lobes, we call the moraine that results from the union of two of these lobes an **interlobate moraine.**

Streams flowing within and beneath a glacier create a variety of interesting landforms. A stream following a meandering path underneath a glacier will deposit its load much as does any surface stream, but when the glacier has receded, the alluvial deposits laid down by the stream will be a sinuously winding rounded mound that may be as much as 50 miles (80 kilometers) in length. These snakelike deposits of neatly stratified alluvium are called **eskers.**

FIG. 15-6 Landforms of Glacial Deposition.

Till Plain

Erratic

Medial, or Interlobate, Moraine

Terminal Moraine

Kame

Recessional Moraine

Kettle

Drumlin

Esker

Alluvium

Outwash Plain

Esker. The esker shown in this aerial photo (a very good example, by the way) winds across the landscape of Boothia Peninsula, Canada's Northwest Territories.

Sometimes earth materials are deposited directly through the glacier, with the motion of the glacier giving the sorted and stratified mound a smoothly rounded shape. Some of these little hills, called **drumlins,** have been found to be almost 1 mile (1.6 kilometers) in length and up to 50 yards (45 meters) thick. Conical hills, or mounds of unsorted and unstratified debris, deposited by a stream running beneath a glacier, are called **kames.** If these deposits are built up along the sides of a glacier, a *kame terrace* will form.

We have already discussed what is probably the most important contribution of glacial deposition to humans: the rich deposits set down by glacial streams and lakes over huge expanses of land that lay beyond the margins of the continental glacier.* Much of the stream-deposited material is silt, and in some places it is estimated to

*Aside from the Great Lakes, North America was given two other lakes of impressive size. One was Lake Bonneville, the ancestor of today's Great Salt Lake, which covered about 20,000 square miles (52,000 square kilometers) and had a depth of close to 1000 feet (300 meters), as compared to today's lake, which measures only 1700 square miles (4420 square kilometers) and is only 30 feet (9 meters) deep. The other was Lake Agassiz, named after the famous glaciologist-zoologist, which once covered much of North America's interior (Manitoba in Canada and Minnesota and North Dakota in the United States), an area of about 100,000 square miles (260,000 square kilometers). Canada's Lake Winnipeg and Lake Manitoba are the only remnants of this massive body of water.

410

be more than 20 feet (6 meters) thick. When the winds pick up and transport this fine silt elsewhere, the resulting finely sifted material (called *loess*) is transformed into a fertile soil, such as that abounding in the Yellow River (Hwang Ho) region of northern China and in much of the United States below the southernmost limit of Pleistocene glaciation.

Present-day continental glaciers and icecaps   All that is left of the continental glaciers that covered so much of the earth is the island of Greenland and the continent of Antarctica.

Antarctica covers an area of 5,100,000 square miles (12,260,000 square kilometers), and its continental glacier covers about 98 percent of the land. In some places, the ice sheet has a thickness of close to 13,000 feet (3900 meters). This considerable mass of ice presses down on portions of the continent, until the actual surface of the land is now below sea level. If, however, the ice sheet were to melt away, the

Antarctica and Its Vast Continental Glacier, Viewed from the *Apollo 17* Spacecraft.

land undoubtedly would rise as the pressure was released. This region is said to be the coldest in all the world, because monthly mean (average) temperatures range from 13°F (−10.6°C) in the summer to −80°F (−62°C) in the winter near the South Pole and are only slightly higher along the coastal regions. The world record for low temperature was set at the Soviet Union's Vostok Station in August 1960: −127°F (−88°C).[4]

In many sections the ice sheet of Antarctica extends out beyond the edge of the continent in great shelves. The Ross ice shelf, for example, is about 1000 feet (300 meters) thick and extends almost 500 miles (800 kilometers) from land. When a section of an ice shelf is broken off and falls into the sea, an **ice floe** is created, a floating island of ice that may be as long as 50 miles (80 kilometers). Can you imagine what seamen of ancient times thought when they first encountered such an iceberg as they rounded the southern tip of South America?

Ice Floe.

Iceberg.

Greenland is the world's largest island. Its area is 840,000 square miles (2,184,000 square kilometers), and of that, 670,000 square miles (1,742,000 square kilometers) is covered with an icecap. It, too, supplies chunks of ice to the oceans, as great tongues of ice push out into the sea and break off (a process called **calving**) and become **icebergs** (ice mountains). While they are not as big or as dramatic as Antarctica's ice floes, they represent a serious threat to the shipping lanes along the western North Atlantic Ocean. Some of these icebergs have been measured to be from 800 to 1000 feet (240 to 300 meters) thick, with most of the bulk hidden beneath the waves. An 800-foot (240-meter) iceberg would have only about 100 feet (30 meters) above the water.

Review and Discussion

1. For what scientific theory is Louis Agassiz noted?

2. Explain why the permafrost regions near the Arctic do not produce glaciers today.

3. Explain the importance of air temperature to the formation of glaciers.

4. Explain the term *snow line*. Why does the snow line vary in altitude at different latitudes?

5. Explain these terms: *zone of accumulation, zone of ablation*.

6. What is meltwater? What is a firn?

7. Explain the difference between these terms: *evaporation, sublimation*.

8. Explain why some glaciers can move more rapidly than others.

9. What is the main difference between alpine and continental glaciers?

10. Define these terms: *bergschrund, abrasion, plucking*.

11. Discuss these terms, which apply to alpine glaciation and its erosional landforms: *tarn, cirque, trough, arête, horn, hanging trough, finger lake, col, roches moutonnées*.

12. What causes striations and glacial polish?

13. As for depositional landforms, what are the following: an erratic, a moraine, a till plain?

14. What is the difference between these moraines: lateral, medial, terminal, recessional, interlobate?

15. What effect did the period of Pleistocene glaciation have on humans?

16. Describe the origin and shapes of the following depositional landforms: esker, drumlin, kame, kettle.

17. What is the difference between icebergs and ice floes?

Notes

[1] Willy Ley (1962), *The Poles* (New York: Time-Life Books), pp. 170–171.

[2] James Gilluly, A. Waters, and A. O. Woodford (1968), *Principles of Geology,* 3d ed. (San Francisco: W. H. Freeman), p. 251.

[3] James L. Dyson (1962), *The World of Ice* (New York: Knopf), p. 5.

[4] U.S. Geological Survey (1969), *The Great Ice Age* (pamphlet 1969 0-357-128), (Washington, D.C.: U.S. Government Printing Office), p. 5.

Additional Reading

Dyson, James L. (1962). *The World of Ice.* New York: Knopf.

Field, W. O. (September 1955). "Glaciers," *Scientific American* (offprint no. 809).

Flint, R. F. (1971). *Glacial and Quaternary Geology.* New York: Wiley.

Ogburn, Charlton, Jr. (1968). *The Forging of Our Continent.* New York: American Heritage.

Shimer, John A. (1959). *This Sculptured Earth: The Landscape of America.* New York: Columbia University Press.

Vial, A. E. L. (1952). *Alpine Glaciers.* London: Batchworth Press.

Woodbury, David O. (1962). *The Great White Mantle.* New York: Viking.

16 Scoured by the Winds

Whether you happen to be standing beside a tent in California's Death Valley, staring at an array of marching sand dunes, or leaning against the driving force of a hot, dry wind on the outskirts of Central Asia's Samarkand, marveling at the seemingly limitless height of a massive dust cloud that reaches to the horizons, you know you are witnessing something that has occurred again and again since time began. You have visible proof of the never-ending labors of the winds.

The Arunta, who live along the northern margins of Australia's Great Sandy Desert, have long been troubled by windstorms that race out of the desert in great whirling clouds of red sand to engulf their encampments and muddy their water holes. To them, it is not a natural phenomenon, because they believe the billowing whirlwinds are the abode of an evil spirit they call Koochee, the Red Demon. When one of these storms appears on the near horizon, the warriors pick up their large boomerangs and rush to the attack.

Shall we consider this Arunta warrior merely another "ignorant and superstitious savage" because he believes he can cut down the whirling cloud of red sand with his expertly thrown boomerang? Perhaps the flat blade of his weapon does, somehow, interrupt the upward flow of spiraling wind, thus robbing the whirlwind of its power.

While it seems doubtful that the roaring column of red sand is inhabited by a demon, it is understandable that the warrior must try to protect himself. His land, the Australian outback, is not the most hospitable place; it is a land of barrenness, of red-pebbled plains, great stretches of shifting sands, and howling winds. If he believes the winds contain spirits, he is no different from hundreds of other peoples living in different places around the world. Winds bring rains, winds bring storms, and winds bring drying conditions and drought. People can harness rivers, dam lakes, or position breakwaters to thwart the ravages of the seas, but as yet, no one has discovered a reliable way to control the winds.

No region on earth is spared the work of the winds. To the nomads of northeastern Siberia, the Yakuts, winds can mean a lowering of warm temperatures during their short summer, or they can mean a chilling blast of icy air from the Arctic that will once again freeze the partially thawed top layer of the permafrost. A century or so ago, the Finns sold wind to masters of becalmed ships. The provided a rope tied in three knots, and if "they undid the first knot, a moderate wind sprang up; if the second, blew half a gale; if the third, a hurricane."[1] Indians of South America fling fiery brands at a strong wind threatening their homes and, in addition, try to further frighten it with their screams.*

The winds never cease their restless whisperings or howlings as they prepare the land for the various weathering processes. They erode the land, transporting particles from one place to another, and they deposit their loads as they too, along with the

---

*For an interesting look at the attempts of mankind throughout history to understand and control the workings of the atmosphere, see: Sir James G. Frazer (1960), *The Golden Bough* (New York: Macmillan), especially chapter 5, pp. 92–96, "The Magical Control of the Wind."

Winds: Rivers of the Lower Atmosphere. In this satellite view of part of southern California, includ-ing the Mojave Desert, a number of mountain ranges, and much of the county of Los Angeles (see the accompanying map), dust that has been roiled by the activities of recreational vehicles is picked up by the winds, resulting in "dust plumes" that have the appearance of flowing streams.

other earth–shaping forces, attempt to bring about a leveling of the land. One might infer that wind currents are somewhat like stream currents and so term them "rivers of the lower atmosphere."

Wind Erosion

While winds operate in a manner similar to streams of running water, they are not confined between the walls of narrow valleys and so can operate freely over wide areas. The erosional processes of wind erosion are abrasion and deflation.

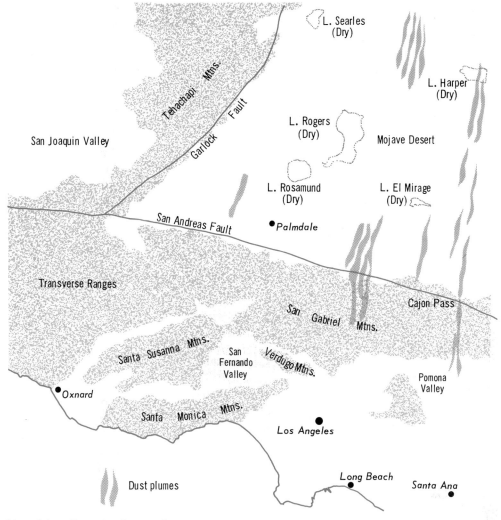

Map of Area Shown in Photograph on Facing Page.

ABRASION We used the term **abrasion** in our discussion of the sculpturing performed by running water, oceans, and glaciers, but the wind's action of wearing away a surface is more akin to "sandblasting"—sand grains being hurled against an object. The process of abrasion does not build landforms; it sculptures those already in existence. The various forces of weathering prepare landforms by breaking apart the rock structures, and then comes the wind, lifting the loosened particles, carrying them along, and throwing them violently against other objects in the path. Because of the weight of sand grains, most of the work of abrasion is accomplished within 6 feet (2 meters) of the ground. If, however, the wind is powerful, moving at a high speed, it can carry a heavier load to higher elevations. While running water performed the major erosional work on the columns of Bryce Canyon National Park, wind abrasion has softened the contours somewhat. Anyone who has been caught in a sandstorm while driving on the desert knows what windblown sand can do to the paint on an automobile. Wind abrasion can also shape large pebbles and rocks into one-, two-, or three-sided shiny fragments, called **ventifacts** (from the Latin, *ventus,* "wind," and *factum,* "done"). Ventifacts usually are very small, with shiny, polished facets, or sides.

DEFLATION   The erosional process of **deflation** refers to the removal of sand or silt particles from an area that has been previously weathered, eroded, or deposited. Deflation occurs mostly in dry (arid or semiarid) climate regions that have little in the way of vegetation to hold onto the loosened particles of a weathered surface. It can also occur, however, in humid regions having flood plains, old lake beds, or the till or outwash plains of glaciated areas, as well as on elevated sea terraces or sea cliffs of an uplifted shoreline.

If the area is composed of loose sand particles, a depression may be created when the small particles are carried away. This hollowed-out space is called a **blowout,** or **deflation basin.** Blowouts vary in size, and although most are small, some attain lengths of up to 1 mile (1 or 2 kilometers). If an area also happens to have a great quantity of gravels and pebbles, the finer particles may be carried away, leaving the larger rock materials behind. This results in a stony, sandless desert floor that sometimes looks much like a paved country road. Geologists call this erosional landform

Desert Pavement in Death Valley, California.

desert pavement, but the Arabs of North Africa's Sahara Desert call it *hamada,* or *reg,* and the Australians call it a *gibber plain*. Despite the movies and television shows that depict the deserts of the world as composed of majestically billowing sand dunes, stony deserts are the most common.

Wind Transportation

As we mentioned earlier, most wind-transported materials are carried along very close to the ground. This is because of the weight of the particles and the velocity of the wind (which is not usually great enough to transport materials heavier than sand or silt). Also, wind is more variable and turbulent than flowing water; thus, motion in a particular direction cannot be sustained by particles being carried along by the wind. Much of the wind's deflation and abrasion is a result of a process called **saltation** (Fig. 16-1). Briefly, saltation is the result of wind-driven sand grains bouncing along the ground, striking other sand particles, dislodging them, and causing them to bounce into the air or to be pushed along slowly with the moving wind. The slow movement of sand grains downwind along the ground is called **surface creep.** If the particles are small enough, dust storms may result, and a considerable amount of soil may be lost to a marginal agricultural region. This happened to parts of Oklahoma and Kansas during the 1930s, when strong winds carried away soil that had been exposed by improper ranching and farming practices, thus creating the famous (or infamous) Dust Bowl.

FIG. 16-1 Saltation: The Movement of Sand Grains by Wind.

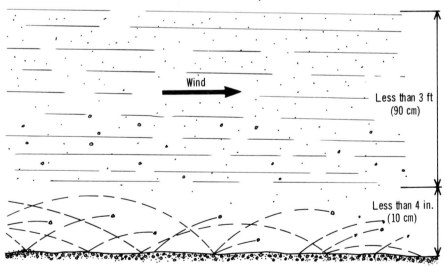

Wind

Less than 3 ft
(90 cm)

Less than 4 in.
(10 cm)

Silt   Sand   Path of grains

Wind Deposition

While winds do perform some significant erosional activities, they are more noted for their depositional accomplishments, such as those resulting from the deposition of sand (sandy deserts and sand dunes) and silt (loess deposits).

SANDY DESERTS If you compare the map in Fig. 7–4, showing the world's deserts, to the map in Fig. 16–2, showing the location of the sandy desert areas, our earlier statement, that stony or pebbly deserts are larger and more common than sandy deserts, will be verified. Stony deserts do have their exciting features, but they cannot match the landforms to be found in the earth's sand regions. It might be noted that the most popular term used to denote a stony desert is the Arabic word *reg,* and for sandy desert, the word is *erg.* It might also be noted that few deserts (if any) are either completely reg or erg. Most have features representing both types. There usually are great stretches of reg, followed by large areas or pockets of erg. Blowouts and desert pavement, along with *wadis* (an Arabic word for gullies that are only intermittently filled with running water), are the outstanding landforms of regs, while sand dunes (and certain related features) are the prominent landforms of ergs.

Sand dunes may be found in any area that has an extensive supply of sand, such as on the fringes of large deserts, along shorelines, in the floodplains of stream valleys in arid climates, or near areas that were once glaciated. The United States has a number of interesting sand–dune areas, such as the Indiana Dunes, Little Sahara State Park in Oklahoma, Sleeping Bear State Park in Michigan, and the Great Sand Dunes National Monument in Colorado. Sand dunes occur in a variety of shapes, depending on the wind direction and velocity, the quantity and size of the sand grains, the amount and type of vegetation, and the topography of the land.

Figure 16–3 details the evolution of a sand dune. Wind drives grains of sand (by saltation) along a surface (Fig. 16–3[a]), until the velocity is slowed by some sort of obstruction (such as a depression in the ground, a small hill or rise, or as in the illustration, some type of vegetation.) The sand grains begin to pile up in the shadow (leeward side) of the shrub (Fig. 16–3[b]), forming a mound composed primarily of the heavier grains (because the lighter grains are transported further on by the wind). As the dune builds (Fig. 16–3[c]), the windward slope rises in a gentle incline (normally about 10 degrees), and the lee side becomes quite steep (about 35 degrees) as the heavy grains can no longer be pushed by the wind. They collect at the top of the dune, then by the pull of gravity, they fall, or slip, down the lee (slip) side. With continuing action (Fig. 16–3[d]), the slip side sharpens, and the dune advances slowly downwind, as sand grains bounce up the windward slope or around the sides of the dune and move on. New layers of sand are added, giving the internal structure of the dune a stratified appearance. If you were to dig into an old dune, you would discover distinct strata, beds of different types of wind-deposited materials. If the wind becomes too strong or replenishment of sand losses is reduced, the dune will diminish in size and, eventually, be reduced to a thin, rippled surface (Fig. 16–3[e]).

FIG. 16–2 Sandy Deserts of the World.

426

FIG. 16-3 Evolution of a Sand Dune.

The most famous of the forms taken by mounds of moving sand is the crescent-shaped dune, called the **barchan**. The crescent shape is caused as sand grains race around the sides of the mound and form "horns" that extend downwind under the directing force of a wind that prevails in one direction and does not shift about.

When the supply of sand and the wind come from the same direction, perhaps at the mouth of a great valley, the sand will be strung out in a long ridge, parallel to the wind's direction, thus forming a **longitudinal dune.** Some of these may extend for 100 miles (160 kilometers), while attaining heights of 330 feet (100 meters).[2] An interesting combination of the longitudinal and barchan shapes occurs when the winds import fairly significant amounts of sand during a season and these deposits are reworked by winds coming from a different direction in a later period. The result is a string of barchans extending along the angle of the dominant wind like links in a chain. This form has been given the name **seif** (an Arabic term meaning "sword.")

If there is a moderate supply of sand, steady winds, and a sturdy vegetative cover, the mound of moving sand may find itself anchored on two sides by large clumps or shrubs, which hold the horns while the main body of sand moves on to take a form that is the reverse of the barchan: The horns point upwind. This is termed the **U-shaped** or **parabolic sand dune.**

Where there is abundant sand and moderate winds, the sandy landscape looks much like the surface of an ocean, as the "waves" appear to march across the land

Barchan Sand Dune. The site of the dune area in which this well-formed barchan was found is about 20 miles (32 kilometers) east of El Paso, Texas.

Sand-Dune March. Constantly shifting sand dunes encroach on the oasis at Fort-Flatters, Algeria, in the Sahara desert region called Erg Edeyen. (FAO Photo)

in an endless procession. The ridges of these dunes, called **transverse** dunes, are at right angles to the wind direction.

If the winds are not constant in their direction, but come from all sides, shifting their directions throughout the day, the result is a **star-shaped** dune. Unlike the other dunes, star dunes remain stationary.

Other wind-deposited landforms include the **whaleback** and the **sand shadow.** Whalebacks are large rounded mounds of sand that may extend for 50 miles (80 kilometers), their axes parallel to the prevailing winds. A sand shadow results when sand collects on the lee side of an object (house, boulder, or shrub) as the eddies of circling wind can no longer support their loads. This action is comparable to the deposits that result when a stream of running water is forced into a backwater eddy on the inside of a stream's meander.

LOESS DEPOSITS **Loess** (pronounced *lurse*) is wind-transported and deposited silt. It is usually composed of fine-grained, silty materials that are buff or grayish in color.

Figure 16–4 shows the location of the major loess deposits of the world, and if you compare this map with the one showing the extent of Pleistocene glaciation (Fig. 15–4), it will become apparent that sediments deposited by streams issuing from the great continental glaciers or from large glacial lakes are the original source of these deposits. Some of the world's most important agricultural areas are in regions where loess deposits are the deepest and enough time has passed to allow vegetation to add humus to what was potentially a fertile soil. The region of loess deposits through which China's Yellow River flows might be termed "China's breadbasket," because this rich, loamy soil helps to support that country's hundreds of millions of people. Argentina's pampa is famous for its cereals, grains, and cattle, as are the fertile steppes of Central Asia and the Ukraine. Other productive loess regions are the North European lowlands of Germany and Poland and, in the United States, the Great Plains, the central lowland, and the lower Mississippi River Valley.

The Imprint of Mankind on a Sandy Desert Landscape.   The strange, footprintlike depressions are date-palm oases in Souf, Algeria (north of Touggourt in the Grand Erg Oriental). The farmers plant palm trees in areas where the water table is relatively close to the surface and erect small fences around the perimeter of each tiny grove. Wind-driven sand (the arrow shows the prevailing direction) billows around the palm oases, building up along the sides. Once past the oases, the sand resumes standard, barchan, shapes.

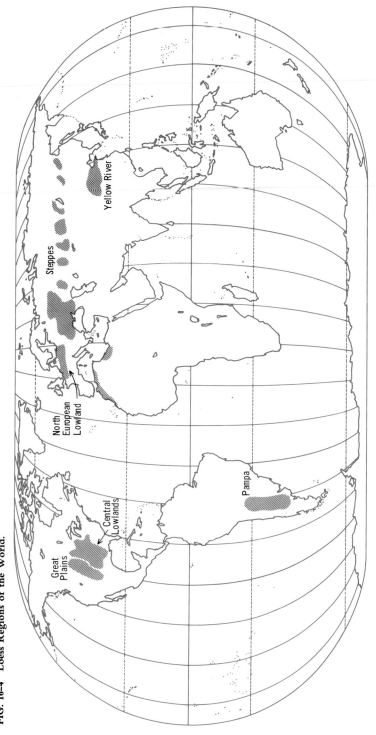

FIG. 16-4 Loess Regions of the World.

Review and Discussion

1. Define these terms, referring to wind erosion: *abrasion, deflation.*

2. Why have winds sometimes been referred to as "rivers of the lower atmosphere"?

3. Describe a ventifact.

4. What is a blowout? What is a deflation basin?

5. What is the difference between deserts called *reg* and *erg*?

6. What is a gibber plain?

7. What is saltation? How does this process differ from surface creep?

8. Why is most abrasion by the wind accomplished close to the ground?

9. What is a wadi?

10. Describe the evolution and construction of a sand dune.

11. What factors determine the size and shape of a sand dune?

12. Describe the appearance and form of construction of these sand dunes: barchan, longitudinal, seif, parabolic or U-shaped, and transverse.

13. What is the difference between a longitudinal dune and a whaleback?

14. How does a sand shadow form?

15. What is loess? Where are the most important loess regions of the world? What is the origin of loess?

Notes

[1] Sir James George Fraser (1960), *The Golden Bough* (New York: Macmillan), p. 93.

[2] William D. Thornbury (1969), *Principles of Geomorphology,* 2d ed. (New York: Wiley), p. 296.

Additional Reading

Bagnold, R. A. (1941). *The Physics of Blown Sand and Desert Dunes.* New York: William Morrow.

Bertin, Leon (1972). *The New Larousse Encyclopedia of the Earth.* Rev. ed. New York: Crown. See especially, "Transport and Erosion by Wind," pp. 46–50, and "Eolian Deposits," pp. 51–53.

Gautier, E. F. (1935). *Sahara: The Great Desert.* New York: Columbia University Press.

Leopold, A. S. (1961). *The Desert.* New York: Time-Life Books.

Lugn, A. L. (1962). *The Origin and Sources of Loess.* Lincoln: University of Nebraska Press.

Epilogue

The Environment of Mankind—Tomorrow?

Two ways the rivers
Leap down to different seas, and as they roll
Grow deep and still, and their majestic presence
Becomes a benefaction to the towns
They visit.

THE GOLDEN LEGEND, LONGFELLOW

It is hoped that this course of study into man's physical environment has made the point that the sun and all of the earth's components (climates, waters, vegetation, soils, and the rocky mass of the earth itself) are interrelated—interconnected by a number of cycles. The earth and its atmosphere, as well as all the other elements, exist together, and a change in any one of them affects all the others. This knowledge, although too often disregarded in the past, is not new. In the second century A.D., one of the more compassionate of the Roman Caesars, Marcus Aurelius, wrote:

> *Regard the universe often as one living being, having one substance and one soul; and observe how all things act with one movement; and how all things cooperate as the causes of all that exists; observe too the continuous spinning of the thread and the single texture of the web.*[1]

Perhaps now, with the acceleration of our knowledge of the processes of nature, we are ready to hear and to understand this message.

We are standing on the threshold of new discoveries—developments that may further increase our understanding of our environment and thus enable us to make decisions that will improve, rather than cause to deteriorate, our quality of life, and preserve, rather than destroy, our world resources.

Whether or not one plans to pursue a career in the field of geography, knowledge of the physical world—its history and its processes—becomes ever more important as man's population and technological capabilities increase.

We have indicated that humans are both users and abusers of the earth. What can professional geographers do about the abuses? What opportunities and challenges are offered to those who would like to participate actively in the advancement of technology without destruction to the environment?* A number of directions can be taken. Some might aspire to solve the immediate problems confronting our neighborhoods, towns, or cities, while others might be more concerned with the long-range effects of human interference in the workings of nature.

Consider the problem of the shortage of petroleum. What should be done? Should we allow offshore drilling on the outer continental shelves? Should we encourage strip mining to reap close-to-the-surface coal deposits? Should we push for immediate development of the huge oil-shale deposits in the western regions of the United States? Should the Canadians strive to retrieve the oil reserves held in the tar sands along the Athabasca River north and east of Edmonton? Are there ways to tap these resources without upsetting the environment?

*See Salvatore J. Natoli, ed. (1974), *Careers in Geography,* rev. ed. (Washington, D.C.: Association of American Geographers), for an interesting survey of the career possibilities in the field of geography.

Perhaps geographers can do more to educate those who might despoil the land or sea, just as soil scientists have aided farmers in learning how to replenish their land by allowing it periodically to lie fallow. The miner or the oil driller cannot replace the oil or the coal (or other minerals) he has taken from the land, but he can learn to restore its former contours and to replace its vegetation by reforestation, thus starting again its ecological cycle.

Geographers might become members of urban and rural planning and redevelopment commissions to encourage measures that would prevent, or at least minimize, pollution and deterioration. They might use their knowledge of climatology and physical geography to aid in deciding where to locate highways and railroads, factories and industrial developments, and new agricultural regions. As resource evaluators (or land-use ecologists, or community development engineers, or economic geographers), geographers might evaluate an area's possibilities with regard to soils, minerals, waters, forests, grasslands, agricultural crops, wildlife, fisheries, and recreation areas. As a physical resource person, he might become concerned with the development of new forms of energy, such as solar, geothermal, and atomic energy. He might map the distribution of the resources of an area and aid in planning for the most beneficial development and utilization of those resources with the least detriment to the environment.

For those who plan to become teachers, the material presented in this course might advantageously be augmented with courses in human or cultural geography to enable you to explain the origins and movements of people and ideas from one place to another throughout the history of civilization. Both you and your students would then see more clearly the reasons why cultures and technologies have become so varied across the face of the earth.

Where can geographers find opportunities for putting their skills and knowledge into action? There are governmental agencies (federal, state, and local) that need well-trained geographers, specialists in the various fields. For example, consider the possibilities in the Department of the Interior's Bureau of Land Management, Bureau of Reclamation, Geological Survey, National Park Service, and Department of Forestry and in the Department of Commerce's Coast and Geodetic Survey (for cartographic work), the U.S. Weather Bureau, the Area Redevelopment Administration, and the Environmental Science Services Administration. Also, there are opportunities in the Soil Conservation Service of the Department of Agriculture.

Businesses need geographers to help them solve environmental problems, such as how to deal with sewage outfall or industrial wastes that pollute streams, lakes, and oceans, or exhausts that pollute the air. Geographers can also aid businesses in developing markets at home and abroad, as well as help to establish transportation routes for materials and products.

For those who are interested in "emerging" countries, there are opportunities with the Agency for International Development (AID), which administers economic aid to foreign nations.

Man's technology has caused many of today's problems, but this same technology can also open doors to intelligent environmental planning. People who develop an awareness of the processes of our environment, whether they work as professional geographers or not, can contribute significantly to the reversal of the heretofore dominant trend to abuse rather than efficiently to use the world's resources.

Note

[1] Irwin Edman, compiler (1945), *Marcus Aurelius and His Times* (Roslyn, N.Y.: Walter J. Black), p. 41. With permission.

Additional Reading

Allsopp, Bruce (1972). *The Garden Earth: The Case for Ecological Morality.* New York: William Morrow.

Darling, F. Fraser, and John Milton, eds. (1966). *Future Environments of North America: Transformation of a Continent.* Garden City, N.Y.: Natural History Press.

Ehrlich, Paul R., and Anne H. Ehrlich (1970). *Population/Resources/Environment: Issues in Human Ecology.* San Francisco: W. H. Freeman.

Johnson, Cecil E., ed. (1972). *Eco-Crisis.* New York: Wiley.

Johnson, Warren A., and John Hardesty (1971). *Economic Growth vs. the Environment.* Belmont, Calif.: Wadsworth.

McKenzie, Garry D., and Russell O. Utgard, eds. (1972). *Man and His Physical Environment: Readings in Environmental Geology.* Minneapolis, Minn: Burgess.

Manners, Ian R., and Marvin W. Mikesell, eds. (1974). *Perspectives on Environment: Essays Requested by the Panel on Environmental Education, Commission on College Geography.* Publication no. 13. Washington, D.C.: Association of American Geographers.

Marsh, George Perkins (1965; originally published in 1864). *Man and Nature.* Ed. David Lowenthal for the John Harvard Library. Cambridge, Mass.: Belknap Press, Harvard University Press.

Odum, Howard T. (1971). *Environment, Power, and Society.* New York: Wiley.

Pursell, Carroll, ed. (1973). *From Conservation to Ecology: The Development of Environmental Concern.* New York: Thomas Y. Crowell.

Strahler, Arthur N., and Alan H. Strahler (1973). *Environmental Geoscience: Interaction between Natural Systems and Man.* Santa Barbara, Calif.: Hamilton.

Appendixes

A Time Zones

When we discussed the earth's rotation and solar (sun) time in chapter 3 (page 60), you may have gotten the idea that it is simple to figure out what time it is somewhere else in the world. Because we know that the earth rotates 15 degrees every hour, it should be merely a problem of determining the number of degrees between the other location and your own. Then, knowing your own time, you should be able to add or subtract the difference to find out what time it is in the other location. For example, if a place is 90 degrees west of your city, there would be six 15-degree units between that place and you. Therefore, the time at the other location would be six hours earlier in the day than in your city. However, such is not the case. The countries of the world do not set up their time zones according to 15-degree intervals. Rather, they choose their time zone to suit their particular interests and problems. Consider Fig. A–1, which shows today's world time zones, and you will immediately realize that the north-south lines separating one time zone from another do not follow meridians in a straight line from north to south. They vary tremendously. For example, check the variations at these places: the Galapagos Islands off Ecuador; Southeast Asia; India and Bhutan, which surrounds Bangladesh; England, as opposed to France; and the People's Republic of China. You will note that the numbers +1, +2, etc., are used to denote the number of hours east of Greenwich, England, and −1, −2, etc., denote the hours west of Greenwich. Note also India's time is +5h 30m, which means that this country is five hours and thirty minutes east of Greenwich. Liberia shows a time zone of −0 45m, which tells us that Liberia is 45 minutes in time west of Greenwich—although that country sits astride the prime meridian.

For the world traveler, whether businessman, college student, or tourist, the vagaries of "local time" can be confusing. One either must ignore his watch or constantly reset it as he flashes across the earth from country to country. In other situations, he might fly thousands of miles without changing his watch at all—as in China. If a person flew out of Shanghai at noon, heading westward from the Pacific Ocean coast, he would fly through three hours of solar time (45 degrees), a distance of some 2500 miles (4000 kilometers),

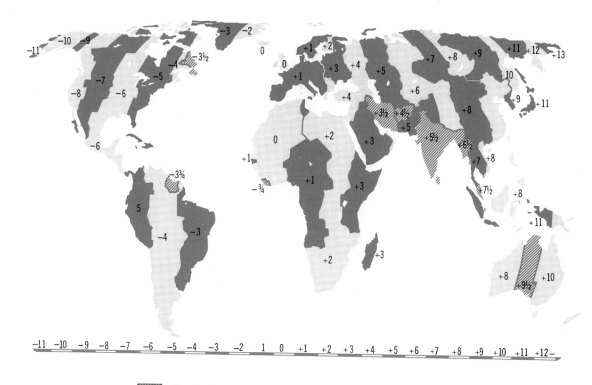

Area that is more or less than one hour different from an adjacent time zone

FIG. A-1   World Time Zones.

before departing China and entering the Soviet Union. He would not, however, alter the time on his watch—unless the plane flew at a speed of 833 miles per hour (1337 kilometers per hour); then, when he reached the border, he would set his watch back to noon.

The zero meridian is the internationally recognized starting point for world time. The prime meridian runs from the North to the South Pole, passing through Greenwich, England. On the opposite side of the world is the 180° meridian, called the International Date Line (Fig. A–2). It is twelve hours distant from the zero meridian: When it is noon in Greenwich, it is midnight along the International Date Line (IDL). If you cross the IDL, traveling from the east toward the west, the day immediately changes. If you left Honolulu on Monday, the minute you crossed the IDL the day would become Tuesday. If it were Monday night at midnight, the day would become Tuesday night, midnight—and you would have lost an entire 24-hour day. If you were traveling from Tokyo to Seattle, and it was Tuesday night, midnight, when you crossed the IDL, it would immediately become Monday night, midnight, and you would live Tuesday all over again. Note also that the IDL zigs back and forth, depending on the needs and wishes of the people residing in each region. See how the IDL bypasses the Aleutian Islands? This is to enable the islands

441

442

FIG. A-2 The International Date Line (IDL). The IDL generally follows the 180° meridian (which is the other half of the great circle that includes the zero, or prime, meridian) and is a line of separation between east and west longitudinal positions. It is deflected occasionally from its straight north-south course to allow certain islands and other regions that are closely related to remain in the same time zone and in the same day.

to be in the same time zone as the state of Alaska. The same situation exists in the Southern Hemisphere, where people having ties with New Zealand want their area to be in the same time zone as that country.

All the problems we have mentioned are repeated within the United States and Canada (Fig. A–3). Rather than following the meridians at 15-degree intervals, the boundaries of the various regions have been allowed to move east or west as the region's needs require. Consider the difference in latitudinal distance between Bismarck, North Dakota, and Montgomery, Alabama—14°29'; this amounts to almost one hour in time, yet both cities are in the same time zone. Both have the same time, although when it is noon (sun time) in Montgomery, it is about 11 A.M. in Bismarck.

FIG. A-3   Time Zones: United States and Canada.

B Topographic Maps

Topographic maps, as all other maps, represent a portion of the earth's surface and can show natural or cultural features. They are laid out on a map projection, usually a modified polyconic projection, which keeps distortion to a minimum.*

However, a well-executed "topo map" is more informative than most other maps. Although any map can depict streams and rivers, mountains and valleys, highways and trails, residential areas, industrial sections, camping spots, farms and other agricultural areas, as well as swamps, beaches, and islands, it cannot do so quite as well as can a topographic map.

Who creates the topographic map? In the United States, it is principally within the province of the U.S. Geological Survey (USGS). The USGS was established by an act of Congress in 1879 to consolidate a varied group of organizations that had been engaged in mapping and the study of the geology and topography of the land, especially public (not privately owned) lands. By 1882, the USGS had developed a general plan to standardize the projected series of topographic maps. Under this plan each map would cover a four-sided area, bounded by meridians of longitude and parallels of latitude; they were, therefore, called *quadrangle maps*.

There are several series of maps, a *map series* being defined as a "family of maps conforming generally to the same specifications or having some common unifying characteristics such as scale. Adjacent maps of the same quadrangle series can generally be combined to form a single large map."†

Specific steps must be followed in order to be certain of accuracy and usefulness, and these are control surveys, mapping procedures, and national standards.

---

*A polyconic projection is one in which many cones intersect the surface of the globe; there is, in fact, a cone for each parallel of need. The meridians are ellipses, with the central meridian of the projection straight up and down. The parallels are not parallel, which would be a serious problem if a large area were to be represented. However, the U.S. Geological Survey uses only a small portion of the projection, so the defects are minimized.

†A little booklet entitled *Topographic Maps* can be obtained from the Map Information Office, U.S. Geological Survey, Washington, D.C. 20242. Indexes showing published topographic maps in each state, Puerto Rico, and the Virgin Islands, are available, free on request, from the USGS (address above), or from the Federal Center, Denver, Colorado 80225.

Geodetic Control Point. Control points, commonly called bench marks, are usually marked on the ground by metal tablets 3¾ inches (9.4 centimeters) in diameter. These marks, placed after measurement, serve for both horizontal and vertical control.

Control surveys are undertaken to locate precise positions of latitude and longitude and height, or elevation. Horizontal control is needed to maintain the correct scale, position, and orientation of the map. The first step, therefore, is to make field surveys. These are conducted by surveyors and engineers, who go out onto the land with accurate instruments to determine the geographic grid and to establish the elevation above sea level of the beginning point. When the "control point" is finally determined, a *bench mark* (or geodetic control point) is fastened to the ground by means of a metal tablet embedded in concrete, rock, or masonry. In the center of this circular piece of brass, the elevation above sea level is stamped.

Mapping procedures have changed considerably since early times. Although the survey teams still go out into the field, they no longer spend as much time there. Today we have aerial photography and photographs from reconnaissance satellites.* Costs to produce highly accurate topographic maps by these methods are about 10 percent of the cost of land-based teams. The process of converting information about elevation (to draw the contour lines) and natural and cultural features from aerial photography is called *photogrammetry*. Here, the cartographer uses stereoscopic viewers to study two photos that have overlapping sections, thus producing a three-dimensional representation of the earth's surface. Sometimes the land-survey party is "too close to the forest to see the trees," and a photograph from space provides startling information, overlooked by those on land. When all the data are accumulated, the map is created. Additional field checking is undertaken, details are added, and finally, place names and political boundaries are entered.

National standards of accuracy in both horizontal and vertical phases of each quadrangle are an important consideration, and each map that meets these standards carries a statement to that effect in the lower margin. For example, horizontal positions of at least 90 percent of the well-defined features must be correctly plotted within one-fiftieth inch on the published map, and elevations of 90 percent of the points tested vertically shall agree with the elevations interplotted from the contour lines within one-half the contour interval. This tolerance allowed for horizontal positions is equivalent to 40 feet (12 meters) on the ground for 1:24,000-scale maps and about 100 feet (30 meters) on the ground for 1:62,500-scale maps.

Map scales represent the relationship between a distance on the map to the actual distance of that part of the earth's surface represented by the map. Topographic map scales are usually presented as a graphic (bar) scale and as a representative fraction scale (Fig. B–1; see also Table 2–2). For example, 1:24,000 means that one unit on the map equals 24 thousand of the same units on the earth's surface. Therefore,

*A booklet published by the USGS Earth Resources Observations System, entitled *Studying the Earth from Space,* is available (35¢) from the U.S. Government Printing Office, Washington, D.C. 20402.

Airborne Radar Imagery: Remote Sensing.   The new scientific field of remote sensing uses electromagnetic equipment and cameras with infrared film. These techniques enable earth scientists to expand their knowledge of the earth far beyond that gained from the use of black and white or color aerial photographs or from study in the field. *Remote sensing* means, literally, "detecting the nature of an object without actually touching it." Remote-sensing equipment carried by high-flying aircraft and orbiting satellites can detect (sense) the differences in infrared reradiation emitted by soils, minerals, vegetation, waters, and man-made features. The images produced by the equipment provide information that cannot be known from closer observation, such as the existence of mineral deposits, the changing salinity of a body of water, the spread of disease in agricultural areas or regions of natural vegetation, and the presence of fractures or faults in the earth's crust. In this remote-sensing photograph of the Salinas River Valley south of Monterey Bay, California (taken from *Skylab 3* space station in earth orbit), the irrigated agricultural area is easily distinguishable, not only because of the highway and rectangular field pattern but also because of the color differences, enhanced by infrared photography.

LARGE
SCALE

1:24,000 scale

1 in. equals 2000 ft.

Area shown is 1 square mile.

MEDIUM
SCALE

1:62,500 scale

1 in. equals nearly 1 mile.

Area shown is 6 square miles.

SMALL
SCALE

1:250,000 scale

1 in. equals nearly 4 mi.

Area shown is 95 square miles.

FIG. B-1 Topographic Map Scales: Large—Medium—Small.

1 inch equals 24,000 inches, or 1 foot equals 24,000 feet, and so on. One can also convert the right-hand number to another unit form, such as converting the R.F. 1:24,000 to a verbal scale, for example, "1 inch equals 2000 feet," or "1 centimeter equals 240 meters." The decision of which scale should be used depends on the amount of detail needed. A map with a large scale is one that shows an extensive amount of detail and is especially useful for highly developed areas or rural areas where detailed information is needed for engineering planning or similar purposes.

Medium-scale maps (1:62,500) may be adequate for rural areas where less detailed planning is contemplated. Small-scale maps (1:250,000 and smaller) cover very large areas on a single sheet and are useful in the study of extensive projects or regional planning. As to which map series should be used by a camper, backpacker, or skier, it is obvious that it would be the largest scale map (Table B-1) available—the 7½-minute series, which covers about 49 to 70 square miles on one sheet.

The USGS is not the only agency producing and publishing maps for our use. Other federal agencies include the U.S. Army Map Service, Tennessee Valley Authority, Mississippi River Commission, and (for the person who enjoys ocean sailing) the U.S. Coast and Geodetic Survey, which publishes navigational charts of the waters off our coasts.

Topographic map symbols are used to explain the map. In a sense, the symbols comprise a "language" for us to read. Figure B-2 illustrates some of the more common symbols, and it must be understood that symbols come in various styles, shapes, and colors, and that each has its own definite meaning. Usually, the contour lines—the lines that connect points of equal elevation above sea level—are colored brown; streams and other water bodies are blue; natural vegetation, green; buildings, black; boundaries are designated by black dashes and dots; and so on.

Contours cannot be found on the earth; they are created, imaginary lines—lines determined by surveyors' measurements in the field or from photogrammetry. A

TABLE B-1 National topographic map series

SERIES	SCALE	1 INCH REPRESENTS	STANDARD QUADRANGLE SIZE (LATITUDE-LONGITUDE)	QUADRANGLE AREA (SQUARE MILES)	PAPER SIZE E–W N–S WIDTH LENGTH (INCHES)
7½-minute	1:24,000	2000 feet	7½ × 7½ min.	49 to 70	22 × 27[a]
Puerto Rico 7½-minute	1:20,000	About 1667 feet	7½ × 7½ min.	71	29½ × 32½
15-minute	1:62,500	Nearly 1 mile	15 × 15 min.	197 to 282	17 × 21[a]
Alaska 1:63,360	1:63,360	1 mile	15 × 20 to 36 min.	207 to 281	18 × 21[b]
U.S. 1:250,000	1:250,000	Nearly 4 miles	1° × 2°[c]	4580 to 8669	34 × 22[d]
U.S. 1:1,000,000	1:1,000,000	Nearly 16 miles	4° × 6°[c]	73,734 to 102,759	27 × 27

[a]South of latitude 31° 7½-minute sheets are 23 × 27 inches; 15-minute sheets are 18 × 21 inches.
[b]South of latitude 62° sheets are 17 × 21 inches.
[c]Maps of Alaska and Hawaii vary from these standards.
[d]North of latitude 42° sheets are 29 × 22 inches. Alaska sheets are 30 × 23 inches.

Primary highway, hard surface	
Secondary highway, hard surface	
Light-duty road, hard or improved surface	
Unimproved road	
Road under construction, alinement known	
Proposed road	
Dual highway, dividing strip 25 feet or less	
Dual highway, dividing strip exceeding 25 feet	
Trail	

Railroad: single track and multiple track	
Railroads in juxtaposition	
Narrow gage: single track and multiple track	
Railroad in street and carline	
Bridge: road and railroad	
Drawbridge: road and railroad	
Footbridge	
Tunnel: road and railroad	
Overpass and underpass	
Small masonry or concrete dam	
Dam with lock	
Dam with road	
Canal with lock	

Buildings (dwelling, place of employment, etc.)	
School, church, and cemetery	Cem
Buildings (barn, warehouse, etc.)	
Power transmission line with located metal tower	
Telephone line, pipeline, etc. (labeled as to type)	
Wells other than water (labeled as to type)	o Oil . . . . . o Gas
Tanks: oil, water, etc. (labeled only if water)	• • ● ⊘ Water
Located or landmark object; windmill	o . . . . ☒
Open pit, mine, or quarry; prospect	× x
Shaft and tunnel entrance	▪ . . . . Y

Horizontal and vertical control station:	
Tablet, spirit level elevation	BM △ 5653
Other recoverable mark, spirit level elevation	△ 5455
Horizontal control station: tablet, vertical angle elevation	VABM △ 9519
Any recoverable mark, vertical angle or checked elevation	△ 3775
Vertical control station: tablet, spirit level elevation	BM × 957
Other recoverable mark, spirit level elevation	× 954
Spot elevation	× 7369 × 7369
Water elevation	670 670

Boundaries: National	
State	
County, parish, municipio	
Civil township, precinct, town, barrio	
Incorporated city, village, town, hamlet	
Reservation, National or State	
Small park, cemetery, airport, etc.	
Land grant	
Township or range line, United States land survey	
Township or range line, approximate location	
Section line, United States land survey	
Section line, approximate location	
Township line, not United States land survey	
Section line, not United States land survey	
Found corner: section and closing	
Boundary monument: land grant and other	
Fence or field line	

Index contour	Intermediate contour	
Supplementary contour	Depression contours	
Fill	Cut	
Levee	Levee with road	
Mine dump	Wash	
Tailings	Tailings pond	
Shifting sand or dunes	Intricate surface	
Sand area	Gravel beach	

Perennial streams	Intermittent streams	
Elevated aqueduct	Aqueduct tunnel	
Water well and spring	Glacier	
Small rapids	Small falls	
Large rapids	Large falls	
Intermittent lake	Dry lake bed	
Foreshore flat	Rock or coral reef	
Sounding, depth curve	Piling or dolphin	
Exposed wreck	Sunken wreck	
Rock, bare or awash; dangerous to navigation		

Marsh (swamp)	Submerged marsh	
Wooded marsh	Mangrove	
Woods or brushwood	Orchard	
Vineyard	Scrub	
Land subject to controlled inundation	Urban area	

FIG. B-2 Topographic Map Symbols.

(a)

(b)

FIG. B-3 Topographic Maps. Topo maps are drawn to represent a specific portion of earth space. In this figure, (a) represents a hilly coastline complete with cliffs, streams, a highway, and a hooked spit. (b) is a simplified topographic map of the same area, with contour lines set at 20-foot (6-meter) intervals.

certain specific interval (space) is established between the contours (Fig. B–3), such as 20 feet, and we can then tell whether the land is smooth and nearly level or rugged, with hills and valleys, by noting just how the contours are laid out. If the contours are located close together, it means a sharp or rapid difference, a rise or fall that is steep and abrupt. If the lines are widely separated, it means a gentle rise or decline. To determine the elevation of a point situated between two contours, you must interpolate, or approximate, the height of that point.

The margins of topographic maps contain much important information, such as the contour interval, the map's scale, meridians of longitude, parallels of latitude, amount of magnetic declination, geographic location of the mapped area, the year the map was constructed and if and when it was revised, the projection used, and the name of the quadrangle. Take a close look at Fig. B–4, which is a portion of the Bright Angel Quadrangle, Grand Canyon National Park, Arizona. Examine the information supplied in the margins, then study the map itself. What is the contour interval? Can you locate permanent and intermittent streams? Which are deep valleys and which are flat-topped mesas? What is the elevation of the El Tovar Hotel? What is the elevation of Phantom Ranch?

Now, compare this map with Fig. B–5. The area encompassed by the map is exactly the same as that in Fig. B–4—but there is a difference, isn't there? This one has *relief shading* (some call it "plastic shading") to make the buttes, mesas, canyons and valleys, and the river stand out more clearly. Which of these maps do you find easier to read and understand? The USGS has developed a new method of aerial photography to produce maps that are quite easy to understand. They are called *orthophotoquads* (Fig. B–6).

Using the topographic map is not difficult, especially after you have learned to read it and know how to use a compass to orient yourself—to relate your actual location on the earth to that point on the map. Once this is done, you can maintain this relationship by lining up your compass with the geographic north and the magnetic north. Then you can take bearings (compass readings) on distant objects (Fig. B–7), and you will always know your location. Geographers normally refer to a place's *azimuth** to denote its direction from the observer. Azimuth is a place's position with respect to (or relative to) a meridian, forming an angle that is expressed in degrees. A compass has a circle of 360 degrees, and geographic north is 0°, due east is 090°, due south is 180°, and due west is 270°. When you are doing fieldwork or are out on a hike, you will use a magnetic compass to determine the azimuths of places or objects around you, but you have to keep in mind another fact: The needle of the magnetic compass does not point toward geographic north. Indeed, the needle can vary considerably from geographic north—perhaps as much as 15 to 30 degrees. This is because the magnetized point of the needle aims for a large deposit of iron near Greenland. You must keep this variation, or declination, constantly in mind. Fortunately, your topographic map tells you what the variation is, so you can make the proper adjustment.

One final note concerning topographic maps: They are in need of frequent revision. The forces of nature, as well as the hand of mankind, can bring about great changes in both the natural and the cultural landscape. Figure B–8 shows why the USGS believes maps must be revised. In this instance, a quiet and secluded ocean-front retreat became a relatively crowded residential community in less than 16 years. To be certain your map is up to date, check the date listed in the lower margin. After all, who knows what "they" may have done to your favorite vacation or camping area since you were last there.

451

---

*The term *azimuth* comes to us from the Arabic words *al* (the) and *zimut* (the way), thus, "the way" or "the direction" to travel.

FIG. B-4 Standard 15-Minute Series Map: Bright Angel Quadrangle, Grand Canyon National Park, Arizona.

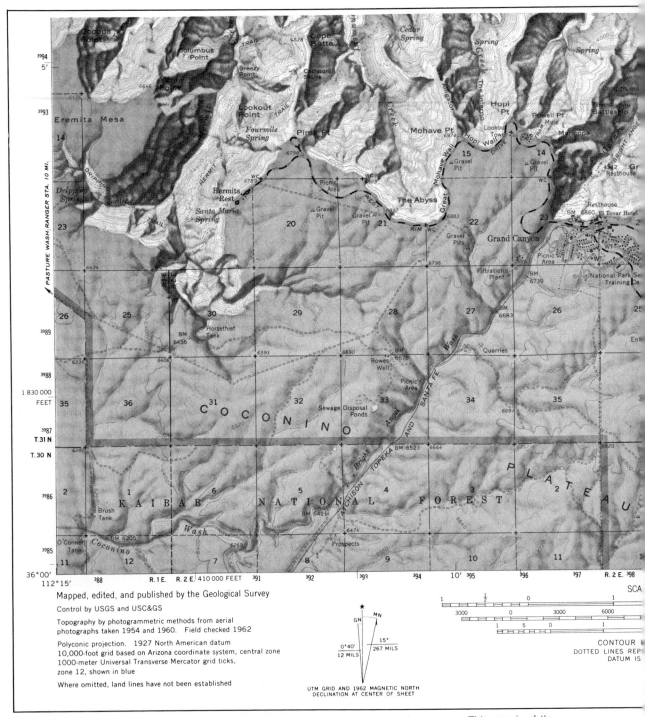

FIG. B-5 Relief-Shaded Map: Bright Angel Quadrangle, Grand Canyon National Park, Arizona. This map is of the me area shown in Fig. B-4. The shading is added to accentuate the unevenness of the terrain.

(a)

(b)

FIG. B-6 Orthophotoquad—Topographic Map. (*a*) is an orthophotoquad (orthophotomap) of a section of Paradise Valley, Arizona. The form of aerial photography illustrated by (*a*) is intended to depict the terrain and other standard topographic map features by color-enhanced photographic images. Compare with (*b*).

(a)

(b)

(c)

FIG. B-7 Direction Finding with a Compass. (a) shows the prime bearings of an object to true north. Due east is 90°, due south is 180°, and due west is 270°. (b) shows a mariner's compass card. Aboard a ship, the ship's course is given as "north by northeast," rather than "steer a course of 020 degrees." (c) shows how to align a magnetic compass with a map's compass rose. Modern maps used for navigation charts or topographic maps also show the amount of magnetic variation, or declination (deviation from true north). For accuracy, geographers are apt to use a 360-degree circle, calling out bearings and azimuths as "azimuth 225 degrees." Adding terminology such as "by north" or "azimuth 225 degrees west of south" is unnecessary.

(a)   (b)

FIG. B-8 Why Maps Must Be Revised. The need for occasional revision of maps becomes obvious when one sees the changes wrought by mankind. (a) shows a portion of Pompano Beach, Florida, in 1946; (b) shows the same area in 1962. Note the inland spread of urbanization, the addition of more canals for marina use, and the major increase in the number of streets and highways.

Additional Reading

Estes, John E., and Leslie W. Senger (1974). *Remote Sensing: Techniques for Environmental Analysis.* Santa Barbara, Calif.: Hamilton.

Holz, Robert K., ed. (1973), *The Surveillant Science: Remote Sensing of the Environment* (Boston, Mass.: Houghton Mifflin).

All map projections we have considered have used the geographic grid to pinpoint the location of landmasses (with the extreme exception of the cartograms), because latitude and longitude is the most efficient and accurate way of determining a place's location. But there are other methods. For example, do you recall reading a story as a child about a pirate who buried treasure he had taken from the ships along the Spanish Main? His map was crude, but the essential points were all there: "Walk away from the beach, until you reach the large black rock; then walk 10 paces north, and turn to your left toward a gnarled tree. Walk 20 paces. Stop. Dig to a depth of 4 feet." For centuries, this was the way people mapped the position of things, and this method was given a name: *metes and bounds.**

Confusion was the ultimate result of the migration of peoples to the New World: Spanish, French, and English, each with their own methods of determining land ownership and boundary lines. For example, when Spanish and Mexican settlers migrated to California, they were given grants of land by the Spanish king, with a description of the particular section of land being granted. However, the description was not tied to any primary or fixed point. When California came under the jurisdiction of the United States government, and its Public Land Survey System was instituted, arguments over the limits of each parcel of land erupted. A new grid system had evolved, with true compass directions and distance measurements more accurately given in feet rather than in steps or paces. Still later, a more systematic method was developed, the rectangular survey system. This method was primarily used for lands being settled in the Midwest and western United States.

In this new land survey system, each plot of land surveyed had definite boundaries, identified by its relationship to a selected meridian (called the *principal meridian*) and a selected parallel (called the *baseline*) (Fig. C–1). The rectangular shapes so arranged are called *townships* (which are 6 miles on each side), and these are subdivided into *sections* (which are 1-mile squares). Figure C–2

*This term comes to us from the original Latin and the later Anglo-Saxon and means: "To measure and fix the limits and the boundary" of a place, without regard to any overall or public plan of survey system.

The Early Spanish Method of Determining Boundaries: The *Diseño* System of Early California. (Adapted from a sketch in W. W. Robinson [1953], *Panorama, A Picture History of Southern California* [Los Angeles: Title Guarantee & Trust Co.], p. 31.)

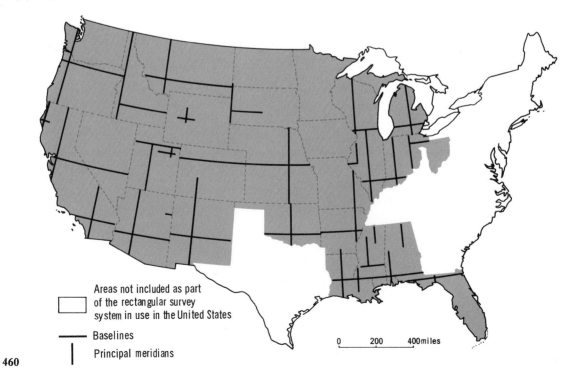

Areas not included as part of the rectangular survey system in use in the United States

Baselines

Principal meridians

0   200   400 miles

FIG. C-2   Township-and-Range System.

Ground Referencing
Systems

461

T4N R4W	T4N R3W	T4N R2W	T4N R1W	T4N R1E

Township Line

T3N R4W	T3N R3W	T3N R2W	T3N R1W	T3N R1E

Township Line

T2N R4W	T2N R3W	T2N R2W	T2N R1W	T2N R1E

Range Line   Range Line   Principal Meridian   Range Line

Township Line

T1N R4W	T1N R3W	T1N R2W	T1N R1W	T1N R1E

Baseline   Standard Parallel

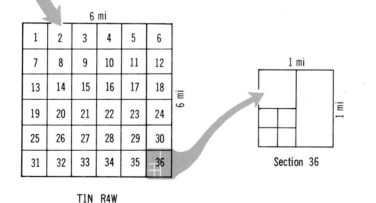

6 mi

1	2	3	4	5	6
7	8	9	10	11	12
13	14	15	16	17	18
19	20	21	22	23	24
25	26	27	28	29	30
31	32	33	34	35	36

6 mi

T1N R4W

1 mi

1 mi

Section 36

FIG. C-1   Rectangular Survey System in the United States.   (on page 460) (For further information on the land survey system used in the United States, see: Francis J. Marschner; U.S. Department of Agriculture [1959], *Land Use and Its Patterns in the United States* [Agricultural Handbook no. 153], [Washington, D.C.: U.S. Government Printing Office].)

illustrates the way the township-and-range system is determined. As you can see, each township has its own designation, such as T4N, R3W, which means "township 4 north (or, fourth township north of the baseline), range 3 west (or, third range west of the principal meridian)." The meridians are always at right angles to the baselines and standard parallels. Each section can be further subdivided—depending on the land-use patterns of the people in that area.

Credits

The author wishes to express his sincere appreciation to the many organizations and individuals who were kind enough to supply photographs for use in this text. Sources for these photographs are listed below, along with sources for the chapter-opening epigraphs. Epigraphs not otherwise credited here or elsewhere were created by the author.

xii-1 National Air Photo Library, Ottawa, Ontario, Canada

Chapter 1

2-3 View of peaks in the Wetterstein Mountains of the Bavarian Alps, West Germany—German Information Center **4** Benjamin Disraeli (1804–1881) quoted from *The International Dictionary of Thoughts,* compiled by John P. Bradley, Leo F. Daniels, and Thomas C. Jones. Copyright © 1969 by J. G. Ferguson Publishing Company; distributed by Doubleday & Company, Inc. With permission. **8** Geodesy, NASA; physical oceanography, Florida News Bureau; astronomy, Lick Observatory Photograph **9** Climatology, Photo Documentation Française; meteorology, U.S. Air Force Photo; hydrology, Tennessee Valley Authority; plant geography, USDA—Soil Conservation Service; pedology, USDA Photo; petrology, Royal Danish Ministry for Foreign Affairs; mineralogy, Geological Survey of Canada #1470; geology, National Park Service; geomorphology, Italian Government Travel Office

Chapter 2

14-15 The west coast of Sweden—Swedish Information Service **18** Musie de Turin **20** NASA **22** NASA **27** Official U.S. Navy Photograph

Chapter 3

54-55 Telephoto view of the rising earth; taken from *Apollo 8*—NASA **56** Pliny the Elder (23–79 A.D.) quoted from *The International Dictionary of Thoughts,* compiled by John P. Bradley, Leo F. Daniels, and Thomas C. Jones. Copyright © 1969 by J. G. Ferguson Publishing Company; distributed by Doubleday & Company, Inc. With permission. **59** JPL/NASA; JPL/NASA **63** Ullberg Photo, Calif. Acad. of Sciences **69** Photo Courtesy of SAS Scandinavian Airlines **71** Photo Courtesy of SAS Scandinavian Airlines **85** EPA

88-89 National Ocean Survey, U.S. Government Photograph

Chapter 4

90-91 Cyclonic storm system; photo taken from *Apollo 9*—NASA **105** NOAA **121** NASA

Chapter 5

Chapter 6

Chapter 7

Chapter 8

Chapter 9

Chapter 10

465

Page references in **bold-faced** type (such as **240**) indicate definitions of a term; in *italic* type (such as *245*) indicate maps, tables, charts, diagrams, or photographs. The letter *n* indicates a footnote.

A

H